# A Shining City on a Higher Hill

## Christianity and the Next New World

The Rt. Rev. James D. Heiser, M. Div., S.T.M.

REPRISTINATION PRESS

First Edition
First Printing: July 2006
Second Printing: June 2012

Repristination Press
P.O. Box 173
Bynum, Texas 76631
www.repristinationpress.com

ISBN 1-891469-49-5

# Table of Contents

# *Introduction*

"What are *you* doing here?" I still smile whenever I remember the question. I was at the 2003 Convention of the Mars Society in Eugene, Oregon and preparing to speak on a panel for one of the evening sessions of the convention when a gentleman I'd never met before at one of our conferences came up to me and asked me his question. "Well," I explained, "I've always been interested in space exploration—especially exploration of Mars." "You know what I mean," he replied, "What are *you* doing here?" Well, I *was* the only person in the room wearing a clerical collar. I endeavored to give him brief reply to his question, and then let my words on that evening's panel serve as a more expanded response.

Such questions don't bother or bewilder me; they are an artifact of a great cultural assumption that 'science' and 'religion' are in 'conflict' and that one must choose sides between 'faith' and 'reason.'

I reject this artificial dichotomy. Faith and reason are both good gifts of the Holy Trinity, but (like all things of this created world) are open to misuse on account of man's fallen nature. In theological circles, one often hears of the distinction between the 'natural' and 'special' revelations—that is, that which the Lord reveals through the creation, and that which He has made known through Holy Scripture. We do not do justice to the truth when we neglect either revelation, but they must be kept in right relationship to one another. Frankly, I believe it is the misunderstanding of these two sources, and of their relationship to one another, which has led to most of the trouble.

What am I doing here? Well, trying to make a contribution to healing the rift which has developed. You may not agree with everything you read in these pages. What I ask of you is that you

**6**

received these essays for what they are: an effort to open a discussion on these points, and an attempt to do justice to the natural and special revelation.

I believe that we stand at a crucial moment in human history; ideas and events are in motion which could have profound significance to the future of our species. Therefore, we need to truly listen to each other—and above all, to listen to the Truth.

*Rt. Rev. James D. Heiser, M.Div. S.T.M.*
*Bishop*
*The Evangelical Lutheran Diocese of North America*

*The Festival of the Presentation of the Augsburg Confession*
*25 June A. D. 2006*

# 1.
# The Keplerian Principle: Space Exploration in a Universe Designed for Discovery.

Few today would question the significance of the changes in Western thought which became manifest roughly five hundred years ago at the beginning of the modern age. Almost every imaginable sphere of human thought and activity has undergone substantial, even fundamental, changes, to the extent that there are many individuals still living who hardly recognize the world of their youth in the world of today. The changes to which we refer are not merely technological; rather, there has been a fundamental alteration in man's perception of the world and his place in it.

One aspect of this change (and it is an important one) is seen in the so-called Copernican Principle. A serviceable description of this principle is as follows:

The Copernican revolution began in the fifteenth century and displaced the Earth as the center of the universe. The Sun, not the Earth, became the center of the solar system. Then the Sun was demoted to but one of many other stars, appearing pointlike and distant in space. By tedious investigation, the stars were later organized into an island universe of hundreds of billions of stars—the Milky Way galaxy. And finally the Milky Way was itself dethroned and became only one of hundreds of billions of galaxies scattered

This essay was originally presented at the 2004 convention of the Mars Society in Chicago, Illinois.

through space. ...

The process of deprovincialization has been very sobering and beneficial to the unchecked human ego, but it has had the unfortunate side effect of seeming to demote the importance of life in the universe. (Malove 51)

Of course, the Copernican Principle borrows its name from Nicholas Copernicus (1473–1543), the famous Roman Catholic priest and astronomer who set forth a heliocentric model of the solar system in his 1543 work, *De Revolutionibus*. Copernicus' model challenged the geocentric model of Ptolemy, which had found nearly universal acceptance in Western thought. It would take nearly a century for scientific observation to validate the Copernican model over the Ptolemaic, but from the outset there were efforts to read philosophical implications into Copernicus' model to which neither he, nor his faithful 'students' such as Michael Mästlin or Johannes Kepler, would have subscribed.

Advocates of the Copernican Principle typically maintain (as was seen above) that Copernicus overturned an egotistical geocentric cosmology, replacing it with a more humble heliocentric one, which in turn, led to a cosmology which viewed Earth as an utterly insignificant world in an unexceptional corner of a unfathomably vast universe. However, there are several problems with such an understanding: first, it mischaracterizes the *humility* of the Ptolemaic cosmology, and, second, it is not faithful to the views of the Copernicans in general, and Kepler in particular. As Gonzalez and Richards observe:

Contrary to popular impression, neither Aristotle nor Ptolemy thought that Earth was a large part of the universe. Aristotle considered it of "no great size" compared with the heavenly spheres, and in Ptolemy's masterwork, the *Almagest*, he says, "The Earth has a ratio of a point to the heavens." ...

More important, the "center" of the universe was considered no place of honor, any more than we think of the center of Earth as being such. And Earth was certainly *not* thought to be "sitting at the center of Heaven itself," as Myrhvold puts it. Quite the opposite. The sublunar domain was the mutable, corruptible, base, and heavy portion of the cosmos. Things were thought to fall to Earth because of their heaviness, and Earth itself was considered the "center" because of its heaviness. The modernist interpretation of geocentrism has it essentially backwards. In our contemporary sense of the words, Earth in Pre-Copernican cosmology was the "bottom" of the universe rather than its "center." (226)

Thus the prevailing medieval cosmology viewed earth as a place far removed from the glories of the heavens; far from being an indicator of "unchecked human ego," it was the Copernicans who believed they were establishing the Earth as a place of honor. Again, to cite Gonzalez and Richards: "So, far from demoting the status of Earth, Copernicus, Galileo, and Kepler saw the new scheme as exalting it. ... He [Galileo] thought that Earth's new position removed it from the place of dishonor it occupied in the Aristotelian universe, and located it in the heavens." (239)

With such an understanding of the philosophy of the historic Copernicans, one immediately recognizes the profound shift which takes place under the Copernican Principle. The Greco-Roman Ptolemaic model, when it was wedded to the Christian worldview, was both humble (since it recognized the relative insignificance of the Earth in relationship to the rest of the universe) and yet still afforded an unmerited grace to humanity (in that the faith upholds that God became man in Christ Jesus to redeem the fallen human race). The modern Copernican Principle maintains that the Earth is an utterly inconsequential place and its variety of inhabitants unremarkable—a view which is increasingly out of step with scientific research. As a result, the Copernican Principle could now be an impediment to sound science, presuming a uniformity which is simply not borne out by the facts. When this principle is joined to materialism, the result is observers who no longer have a philosophical basis for assuming such scientific axioms as causation, lineal time, and the belief that nature is contingent."The official story gives the false impression that Copernicus started a trend, so that removing Earth from the 'center' of the universe led finally, logically, and inevitably to the scientific establishment of our insignificance. By sleight of hand, it transformed a series of empirical discoveries with ambiguous metaphysical implications into the Grand Narrative of Naturalism." (Gonzalez and Richards 244) What is needed is a return to the spirit of early Copernicans such as Johannes Kepler, whose scientific research was guided by the belief that there is a harmony to all of nature, and that man was created in such a way that he could meaningfully study the natural world.

A fundamental assumption underlying science is that the human mind can meaningfully model the natural world. This assumption is ultimately inexplicable for materialism, because there is no need for the *homo sapiens* mind to be capable of the abstract thought necessary to grasp that which is not readily observable. (Gravity might be comprehensible, but quantum theory...?)

The Renaissance assumption of the mind's capacity for understanding the natural world sprang directly from a theological base—a fact readily witnessed in the work of Johannes Kepler. As James Connor recently observed:"Like [Philip] Melanchthon, Kepler expected that the human mind and God's mind worked in roughly the same way, because God created humans in God's own image and likeness." (Connor 67) Again:

> For Kepler, God had planted truth in nature to act as a kind of wordless Scripture, a companion to the Bible. He saw himself, the priest of the Book of Nature, deciphering that book much as a minister of the church deciphers the book of Scripture. ... Writing to [Michael] Mästlin in 1595, he claimed that astronomy should be a practice done in service of God, to give God glory and honor. Such thinking was not only in line with Mästlin's, but also with Philip

Melanchthon's as well.

> Following Melanchthon, Kepler argued that astronomy is natural to humanity, as natural as singing is to songbirds. (Connor 91)

Of course, Connor's comparison of Kepler to Melanchthon makes more sense if one remembers who Melanchthon was.

Philip Melanchthon, who is still revered as the *praeceptor Germaniae* ("teacher of Germany"), was among the most important religious reformers and humanist scholars of the sixteenth century. In a 1536 oration on astronomy and geography, Melanchthon observed:

> As it is most befitting in all things to start with God, so, in this consideration of studies, we should be reminded of the Architect when we contemplate heaven itself. We should consider that He neither established this marvelous order in vain nor handed down the knowledge of these movements to humankind without reason. If someone does not think about this, how does he differ from a Cyclops who, spending his life in a cave, neither looks up to heaven nor worships any eternal god? However, if someone, as it is proper for human nature, makes much of God's gifts he must declare that the laws of these motions were established, and the knowledge handed down, for the sake of great benefits. (Melanchthon 115)

Kepler wholeheartedly embraced such an understanding of scientific study; after all, he was a student not only of mathematics, but especially of theology during his time at the University of Tübingen. As Kepler declared:

> Thus it is apparent that it was not proper for man, the inhabitant of this universe and its destined observer, to live in its inwards as though he were in a sealed room. Under those conditions he would never have succeeded in contemplating the heavenly bodies, which are so remote. On the contrary, by the annual revolution of the earth, his homestead, he is

whirled about and transported in this most ample edifice, so that he can examine and with utmost accuracy measure the individual members of the house. Something of this same sort is imitated by the art of geometry in measuring inaccessible objects. For unless the surveyor moves from one location to another, and takes his bearing at both places, he cannot achieve the desired measurement. (cited in Gonzalez and Richards 238)

Kepler saw a divine intention that man would study the natural world. In his *Epitome of Copernican Astronomy*, Kepler repeatedly refers to Earth as "the home of the speculative creature" and argues that the very structure of the solar system aids in man's study of the universe, and that this is by divine intention.

This understanding of the nature of the universe directly shaped Kepler's approach to astronomy. To cite a fairly typical example, in answering the question, "Then by what means was the magnitude of the Earth adjusted to the solar globe?," Kepler replies:

By means of vision of the sun. For the Earth was going to be the home of the speculative creature, and for his sake the universe and world have been made. But now speculation has its origin in the vision of the stars: wherefore too the magnitude of the things to be contemplated had to have its origin in the magnitude of the things to be seen. But the first visible is light or the sun, as it is (1) the work of the first day and (2) the most excellent of all visible things, the principal, the primary, and that which was going to be the cause of the visibility of all the rest. So it follows that the beginning [*principium*] in proportioning the bodies of the world was taken from the vision of the sun from the Earth; just as in the case of the upper planets the regions of the world were separated by the mean proportionality of the orbit of the Earth. (*Epitome* 33)

Kepler assumed the intelligibility of the universe; he operated on the theological assumption (one which is still fundamental to science) that the human mind had been created in such a way that human observation can provide useful information concerning the natural world. Far from conceiving science and theology to be in tension, Kepler's science flowed naturally from his faith.

Some modern scientists are beginning to operate from what could be called the 'Keplerian Principle': believing that humanity was created with an inherent ability to study profitably the natural world, and that the universe has been created in such a way that those worlds which are habitable are also ideal for scientific study. This is the thesis of Gonzalez and Richards' *The Privileged Planet*. As they observe:

> First, we aren't arguing that every condition for measurability is uniquely and *individually* optimized on Earth's surface. Nor are we saying that it's always easy to measure and make scientific discoveries. Our claim is that Earth's conditions allow for a stunning diversity of measurements, from cosmology and galactic astronomy to stellar astrophysics and geophysics; they allow for this rich diversity of measurement much more so than if Earth were ideally suited for, say, just one of these sorts of measurement. ...
>
> Even more mysterious than the fact that our location is so congenial to diverse measurement and discovery is that these same conditions appear to correlate with habitability. This is strange, because there's no obvious reason to assume that the very same rare properties that allow for our existence would also provide the best overall setting to make discoveries about the world around us. We don't think this is merely coincidental. It cries out for another explanation, an explanation that suggests there's more to the cosmos than we have been willing to entertain or even imagine. (xiii, xv)

**14**

Thus we see that the advance of scientific knowledge is on the verge of bringing us full circle. The philosophical principle known as the Copernican Principle, which has been linked to materialist dogma, rather than being a useful tool, may now be an impediment to science by limiting the vision of scientists. As physicist Stephen Barr recently noted regarding materialism:

> Now, while religious dogmas do not in fact limit the kinds of things one is able to think about, materialism obviously does. The materialist will not allow himself to contemplate the possibility that anything whatever might exist that is not completely describable by physics. That is simply a forbidden thought. It is usually not even felt to be necessary to argue against it. Admittedly, many materialists will say that forbidding one to speak of non-material entities is simply a matter of scientific "methodology." Natural science investigates matter, they say, and so anything that might go beyond matter is outside of scientific discussion. However, it is hard to see why this should be so. For example, one can imagine investigating human psychology in a perfectly scientific way without prejudging whether the human mind is entirely explicable in terms of material processes. In any event, for most materialists it is not really only a question of methodology. The non-material is considered simply beyond the pale of rational discourse. In short, the materialist's notion of what a dogma is, though quite unfair to religious dogma, exactly fits his own views. (Barr 15)

The intellectual perversity of such materialist dogmatism can be seen by considering the fate of Johannes Kepler on a modern university campus. Kepler's contributions to astronomy were *at least* as substantial as those made by Galileo Galilei, and yet he in largely ignored today because of the explicitness of his theological views. The situation is well-summarized by Rowland:

The moral of the story is that Kepler pursued his investigations of the solar system from a position that today's science considers completely off the wall—Pythagorean mysticism, astrology, numerology, Christianity. And yet he comes up with the right answers when Galileo, the father of modern science, could not.

Which demonstrates that there is more than one valid way to do useful science. You don't have to be a mathematical realist—a positivist like Galileo and the rest of his tribe—to do good science. You don't have to buy into the scientistic worldview. In fact, sometimes it can help if you don't.

...that's the reason why Galileo is a hero and nobody's ever heard of Kepler. Kepler is a scientific heretic. His views are heterodox. So he's been suppressed, marginalized. If he were alive today he'd be drummed out of the profession, cut off from grants, denied a teaching position, out of a job... (Rowland 182–3)

If Rowland's assessment seems like hyperbole, one would do well to remember cases such as that of Dr. William Dembski, one of the major proponents of the "Intelligent Design" thesis, who many people believe was academically 'lynched' because of his writings on information theory and evidence for design. If our models have begun to blind us to the truth—if they can no longer accommodate the accumulating body of evidence—then it is time for them to be revised or discarded. As historian John Lukacs recently observed:

Five hundred years ago the Copernican/Keplerian/Galilean/Cartesian/Newtonian discovery—and it *was* a real discovery, a real invention, a calculable and demonstrable and provable one—removed men and the earth from the center of the universe. And often with good intentions. Thereafter, with the exponential growth of scientism, and especially with the construction of ever more powerful instruments, among them telescopes (instruments separating ourselves ever more from what we can see with our naked eyes—but of course the human eye is *never* really "naked"), this movement led to our, and to our earth's having become hardly more (indeed, even less) than a speck of dust on the edge of an enormous dust-bin of the universe, with the solar system itself being nothing more than one tiniest whirl among innumerable galaxies. So many scientists—and not only scientists!—assert this, not at all humbly, but with false intellectual pride. But the discovery in the twentieth century, that the human observer cannot be separated from the things he observes (especially when it comes to his, unavoidably interfering, observation of the smallest component of matter) reverses this. We, and the earth on and in which we live, are back at the center of a universe, which is—unavoidably—an anthropocentric and geocentric one.

But this is something more (and less) than the returning movement of a pendulum. The pendulum of history (and our knowledge of the world) never swings back. It is due to our present historical and mental condition that we must recognize, and proceed from *not at all a proud but from a very chastened view of ourselves*, of our situation, and of our thinking—at the center of *our* universe. (Lukacs 206–7)

If the universe of human experience is now to include Mars, if humanity is going to explore a new world, and even expand its civilization to a new world, then we must do so with, one might say, both eyes open. Human activity is not meaningless; humanity was created for a reason. Many of the great minds of the last age knew this, and our generation must come to believe it once again if we are to continue to build upon their heritage. We must acknowledge that we are not some type of cosmic accident, and that the existence of a world within our reach to explore and possibly settle is not simply one more happy coincidence in an ever more implausible string of 'coincidences.' Only when we free our minds to recognize what the evidence keeps telling us will we be ready to advance in pursuit of our true purpose.

# II.
# Designed for Discovery—
# Is Space Exploration Part of Humanity's Destiny?

O ne does not have to have a great familiarity with the Scriptures of the Old Testament to know about Job. Job's name has become almost synonymous with suffering—to say that someone is undergoing the 'trials of Job' usually doesn't need much explanation: both the one who makes such an assertion, and the one who hears it, understand that the point which is being made is that the person who is so described is suffering immensely, and for no reason which seems immediately apparent to the sufferer. Of course, Job's friends were confident they had correctly analyzed the problem: Job, they thought, had clearly sinned against the Lord God and was being justly punished for his sin; the solution would be for him to repent. However, Job was clearly convinced that their assessment was incorrect, and adamantly professed his innocence. Finally the Lord intervened, presenting Job with a series of questions which reveal the depth of Job's (and his friends') ignorance of the Lord's works and His intentions; e.g., "Can you bind the cluster of the Pleiades, or loose the belt of Orion? Can you bring out Mazzaroth in its season? Or can you guide the Great Bear with its cubs? Do you know the ordinances of the heavens? Can you set their dominion over the earth?" ( Job 38:31–33) In the end, Job says to the Lord, "I know that You can do everything, and that no purpose of Yours can be with-

This essay was originally presented to the 2005 convention of the Mars Society in Boulder, CO.

held from You. You asked, 'Who is this who hides counsel without knowledge?' Therefore I have uttered what I did not understand, things too wonderful for me, which I did not know. Listen, please, and let me speak; You said, 'I will question you, and you shall answer Me.' I have heard of You by the hearing of the ear, but now my eye sees You. Therefore I abhor myself, and repent in dust and ashes." (42:2–7) The Lord's answer to Job wasn't what Job or his friends expected; I suspect it is not what most readers expect the first time they encounter it. The Lord tells Job and his friends that their knowledge of His ways and plans is truly quite limited. Although we can learn something of His ways and His works, the circumstances remain the same: greater knowledge simply continues to highlight the limits of our knowledge—as the horizons continue to recede, the sheer marvel of the creation grows. What we may know with confidence concerning the Lord Himself is only what He chooses to reveal concerning Himself.

For some, in the face of the simple scope of it, they conclude that all of creation must be chaos, that there is not a Creator who has brought all of this into being; Job's friends call upon him to repent, for he must certainly have been mistaken in his claims about the Lord.

However, for others, the same scope, same conditions, same information is further proof that all lends further evidence to order and intention—to design—within the creation, and Job suffers on, desiring that the Lord would vindicate him.

The circumstances of this situation demonstrate the wisdom of a recent observation by George Coyne: "The fundamental problem with all attempts to use the rational processes of science to either assert or deny the existence of God or to limit his action is that they primarily view God as Explanation. We know from Scripture and from tradition that God revealed himself as one who pours out himself in love and not as one who explains things. God is primarily love." (Coyne 181) For the believer, the central emphasis is on his relationship with the Creator; that relationship influences his expectations in his study of the Creator's works, just as the non-believer's lack of such a belief colors his own study, even if he is not consciously aware of such an influence. Again, Coyne notes:

> The Judeo-Christian experience affirms emphatically the enfleshment of the divine and, since God is the source of the meaning of all things, that meaning too becomes incarnate.

> As noted, some see in this religious belief the foundations of modern science. A rigorous attempt to observe the universe in a systematic way and to analyze those observations by rational processes, principally using mathematics, will be rewarded with understanding because the rational structure is there in the universe to be discovered by human ingenuity. Since God has come among us in his Son, we can discover the meaning of the universe, at least it is worth the struggle to do so, by living intelligently in the universe. Religious experience thus provides the inspiration for scientific investigation.

> What are we to make of these assertions? Have we succumbed to a too facile assimilation of religious and scientific experiences? Or, on the other hand, is there truly at the origins of modern science the religious inspiration that God and his plan for the universe are incarnate? At a minimum, these two experiences are not incompatible; and the history of religions and of the origins of modern science certainly appear to support the connection we have presented. (Coyne 184)

Finally, Coyne notes a danger for believers and non-believers alike:

> We must beware of a serious temptation of the cosmologists. Within their culture, God is essentially, if not exclusively, seen as an explanation and not as a person. God is the ideal mathematical structure, the theory of everything. God is Mind. It must remain a firm tenet of the reflecting religious person that God is more than that and that God's revelation of himself in time is more than a communication of

information. Even if we discover the Mind of God, we will not have necessarily found God. (Coyne 186)

Since at least the nineteenth century, Job's friends have increasingly presented themselves as the voice of science and reason, calling upon Job to repent of his notion of a Creator. In the words of Gonzalez and Richards, "Since the nineteenth century, we have been discouraged from asking whether there could be evidence of purpose or intelligent design in nature." (Gonzalez and Richards 332) The argument has often been put forth with a degree of subtlety, essentially attempting to define any non-materialistic explanations outside the pale of discourse. Now, defining the rules of the game so as to automatically exclude your opponent's victory is a great way to play, if you can get away with it. But Job, sensing the metaphysical trick that has been played on him, has gotten increasingly direct in his counter arguments. Thus, for example physicist Stephen Barr of the University of Delaware recently noted regarding materialism:

> Now, while religious dogmas do not in fact limit the kinds of things one is able to think about, materialism obviously does. The materialist will not allow himself to contemplate the possibility that anything whatever might exist that is not completely describable by physics. That is simply a forbidden thought. ... The non-material is considered simply beyond the pale of rational discourse. In short, the materialist's notion of what a dogma is, though quite unfair to religious dogma, exactly fits his own views. (Barr 15)

Indeed, a significant reason for Barr's recent book, *Modern Physics and Ancient Faith* being written was, in his words, to "examine some of the arguments for materialism" for "we shall see that ultimately all of them are completely circular. They all seem to boil down in the end to 'materialism is true, because materialism *must* be true.' The fact seems to be that the philosophy of materialism is completely fideistic in character." (16) Barr's assessment is correct—a fact readily testified to by the emotional reaction which many materialists have when 'called' on this point. However, in *The Privileged Planet*, Guillermo Gonzalez and Jay Richards set out in a different direction. Their studies (Gonzalez's in astronomy, Richards' in theology and philosophy) led them to a new insight in the design model: the realization that those areas of the cosmos most amenable to life were also those which were best for scientific study:

> Simply stated, the conditions allowing for intelligent life on Earth also make our planet strangely well suited for viewing and analyzing the universe.
>
> The fact that our atmosphere is clear; that our moon is just the right size and distance from Earth, and that its gravity stabilizes Earth's rotation; that our position in our galaxy is just so; that our sun is its precise mass and composition—all of these facts and many more not only are necessary for Earth's habitability but also have been surprisingly crucial to the discovery and measurement of the universe by scientists. Mankind is unusually well positioned to decipher the cosmos. Were we merely lucky in this regard? Scrutinize the universe with the best tools of modern science and you'll find that a place with the proper conditions for intelligent life will also afford its inhabitants an exceptionally clear view of the universe. Such so-called habitable zones are rare in the universe, and even these may be devoid of life. But if there is another civilization out there, it will also enjoy a clear vantage point for searching the cosmos, and maybe even for finding us. (xi)

For some, the cosmos exists for no purpose and submits to no explanation more ultimate than itself. For others the cosmos finds its proper interpretation in terms of purpose, design, and intention.

> ... For those who already believe than nature exists for a purpose, perhaps as a result of a more specific set of

philosophical or religious views, our argument may be satisfying, even expected. The skeptic may rightly point out that such individuals may be imposing an artificial pattern on the evidence, much like psychiatric patients impose patterns on Rorschach blots. But such skepticism is a double-edged sword, since such "skeptics" may have blinded themselves to the existence of real patterns in the natural world. The reasonable skeptic—as opposed to the hardened skeptic for whom no evidence is sufficient—should at least consider the possibility that nature exists for a purpose. For those open to such a possibility, the correlation between habitability and measurability should be a compelling discovery. (Gonzalez and Richards 331–2)

It is the very intelligibility of the natural world to human senses and minds—indeed, even regarding those parts of creation beyond the natural observation of our physical senses—which testifies to a Creator, especially since those areas of the creation generally best suited for science are the habitable zones. Again, in the words of Gonzalez and Richards:

Most scientists presupposed the measurability of the physical realm: it's measurable because scientists have found ways to measure it. Read any book on the history of scientific discovery and you'll find magnificent tales of human ingenuity, persistence, and dumb luck. What you probably won't see is any discussion of the conditions necessary for such feats, conditions so improbably fine-tuned to allow scientific discoveries that they beg for a better explanation than mere chance.

... [W]e aren't arguing that every condition for measurability is uniquely and *individually* optimized on Earth's surface. Nor are we saying that it's always easy to measure and make scientific discoveries. Our claim is that Earth's conditions allow for a stunning diversity of measurements, from cosmology and galactic astronomy to stellar astrophysics and geophysics; they allow for this rich diversity of measurement much more so than if Earth were ideally suited for, say, just one of these sorts of measurements. ...

> Even more mysterious than the fact that our location is so congenial to diverse measurement and discovery is that these same conditions appear to correlate with habitability. This is strange, because there's no obvious reason to assume that the very same rare properties that allow for our existence would also provide the best overall setting to make discoveries about the world around us. We don't think this is merely coincidental. It cries out for another explanation, an explanation that suggests there's more to the cosmos that we have been willing to entertain or even imagine. (Gonzalez and Richards xiii, xv)

The authors of *The Privileged Planet* maintain that such an approach to the study of creation leads scientists to anticipate and seek out connections and intelligibility. Often most scientists do not realize how fundamentally their entire approach to the natural world is framed by a Christian world view. It should be pointed out, for example, that "Ockham's Razor" (attributed to the Franciscan theologian, William of Ockham) is a profoundly theological argument rooted in a Christian understanding of the Creator. What one believes concerning the Author of creation directly influences the way in which one seeks order and intelligibility within the creation. The axiomatic "Ockham's Razor" cannot, of course, be proven by scientific study; rather, it is more fundamental, providing an element of the intellectual framework within which scientific study can be done.

> When scientists read nature accurately, nature discloses itself in new and unanticipated ways, like a rich and multifaceted text to the patient interpreter. A proper reading creates new lines of research and exploration. Herein lies a virtue in see-

ing the correlation between habitability and measurability as the result of purpose rather than mere coincidence: we should expect to find it elsewhere, and we should expect to continue making discoveries because of it. To one who has discerned that the cosmos is designed, this correlation is much like the sublime beauty and the mathematical elegance of the natural world—no longer a troublesome anomaly to be explained away but something simultaneously fitting and wonderful. Viewing it as a mere coincidence, in contrast, is both theoretically and aesthetically sterile. (Gonzalez and Richards 333)

Such an understanding of the creation clearly predominated at the very time when the foundations of the modern sciences were being strengthened. Philip Melanchthon, the 'teacher of Germany,' observed in 1531:

Who is so hard-hearted and so without feeling that he does not sometimes, looking up at the sky and beholding the most beautiful stars in it, marvel at these varied alternations which are produced by their motions, or desire to know the traces, so to speak, of their motions, that is, the fixed computation shown by divine providence? For such varied things, which are placed so far away, would not have been investigated or perceived by human sight had God not roused and advanced the studies of some outstanding men. Since therefore nature leads men to these arts, one has to consider as utterly lacking a human mind those who are in no way affected by the sweetness of these things and of knowledge. (106)

Or as Johannes Kepler noted regarding the Copernican understanding of the solar system:

Thus it is apparent that it was not proper for man, the inhabitant of this universe and its destined observer, to live in its inwards as though he were in a sealed rome. Under those conditions he would never have succeeded in contemplating the heavenly bodies, which are so remote. On the contrary, by the annual revolution of the earth, his homestead, he is whirled about and transported in this most ample edifice, so that he can examine and with utmost accuracy measure the individual members of the house. (quoted in Gonzalez and Richards 238)

For an example of a modern scientist operating with such an understanding of the creation, we need look no further than Dr. Wernher von Braun, the father of the U.S. space program. Throughout his life, von Braun's faith was intimately connected to his scientific interests. Even at the time of his Lutheran confirmation as a boy, his mother gave him a telescope as a confirmation gift!

Dr. von Braun's belief extended to a firm confidence that the Lord of the creation heard and answers prayer. As he readily declared, "I certainly prayed a lot before and during the crucial Apollo flights, and I also prayed during the last days in Germany—when things collapsed all around me." (Bergaust 117) Or one might consider the account which Ward relates in his recent biography of von Braun:

A different side of von Braun showed in the case of Kevin Steen, a boy from Carefree, Arizona, who had written in the late 1960s to von Braun in Huntsville about his enthusiasm for space flight. Kevin developed cancer at an early age. Von Braun responded with a letter of encouragement, and soon the two became pen pals. Not long after the scientist had transferred to Washington his secretary received a call from Kevin's father. He wanted her to tell von Braun that the boy had undergone surgery at the Mayo Clinic but that "his body was filled with cancer and the surgeons could do nothing," Kertes recalled.

"Dr. von Braun said [to his staff], 'We must all pray for Kevin,' and we did," the secretary said. "To everyone's amazement, a miracle happened and Kevin began to recuperate. A few months later we learned that Kevin had been back to Mayo and there was no sign of cancer in his body!" The secretary remembered well her boss's comment to the staff: "Now we can see what the power of prayer can do." (189)

Through his professional career, von Braun was outspoken concerning the importance of recognizing design within the creation. In a letter to the California State Board of Education (which he wrote in support of teaching both the materialist and 'design' models) Dr. von Braun stated:

> For me, the idea of a creation is not conceivable without invoking the necessity of design. One cannot be exposed to the law and order of the universe without concluding that there must be design and purpose behind it all. In the world around us, we can behold the obvious manifestations of an ordered structured plan or design. We can see the will of the species to live and propagate. And we are humbled by the powerful forces at work on a galactic scale, and the purposeful orderliness of nature that endows a tiny and ungainly seed with the ability to develop into a beautiful flower. The better we understand the intricacies of the universe and all it harbors, the more reason we have found to marvel at the inherent design upon which it is all based.
>
> While the admission of a design for the universe ultimately raises the question of a Designer (a subject outside of science), the scientific method does not allow us to exclude data which lead to the conclusion that the universe, life and man are based on design. To be forced to believe only one conclusion—that everything in the universe happened by chance—would violate the very objectivity of science itself. Certainly there are those who argue that the universe evolved out of a random process, but what random process could

produce the brain of a man or the system of the human eye?
(quote in Bergman)

In fact, Dr. von Braun marveled at the attitude of those scientists
who were unwilling to weigh seriously the possibility of a Designer.

> I have discussed the aspect of a Designer at some length
> because it might be that the primary resistance to acknowl-
> edging the "Case for Design" as a viable scientific alternative
> to the current "Case for Chance" lies in the inconceivability of
> some ultimate issue (which will always lie outside scientific
> resolution) should not be allowed to rule out any theory
> that explains the interrelationship of observed data and is
> useful for prediction.

> Many men who are intelligent and of good faith
> say they cannot visualize a Designer. Well, can a physicist
> visualize an electron? The electron is materially inconceivable
> and yet, it is so perfectly known through its effects that we
> use it to illuminate our cities, guide our airliners... and take
> the most accurate measurements. What strange rationale
> makes some physicists accept the inconceivable electron as
> real while refusing to accept the reality of a Designer on the
> ground that they cannot conceive Him? I am afraid that,
> although they really do not understand the electron either,
> they are ready to accept it because they managed to produce a
> rather clumsy mechanical model of it borrowed from rather
> limited experience in other fields. (quoted in Bergman)

Von Braun's last major writing was done for a convention of the
Lutheran Church in America. Ward comments concerning that
1976 work:

> As for the true impetus for space travel and ex-
> ploration, the rocket pioneer stated he was convinced that
> "the answer lies rooted not in whimsy but in the nature of
> man. I guess it is all just in the basic makeup of man as God

wanted him to be. I happen to be convinced that man's newly acquired capability to travel through outer space provides us with a way out of our evolutionary dead alley."

His summing-up before the Lutheran gathering also reflected his tightening embrace of his own Christian faith as he faced death. He revisited what had become his familiar rationale that science and religion are fully compatible. "In this reaching of the new millennium through faith in the words of Jesus Christ, science can be a valuable tool rather than an impediment. The universe revealed through scientific inquiry is the living witness that God has indeed been at work. Understanding the nature of the creation provides a substantive basis for the faith by which we attempt to know the nature of the Creator." (Ward 217–8)

In this faith, von Braun lived and worked, and in this faith he died. On his tombstone are carved the simple words: "Wernher von Braun, 1912–1977, Psalm 19:1." That passage of Holy Scripture declares: "The heavens declare the glory of God, and the firmament showeth His handiwork."

If Kepler, von Braun, and others are right, there is merit to understanding the unity of creation which manifests its origin in the work of the Creator. And certainly Kepler and von Braun would have agreed with the substance of Coyne's concern: that the Lord is not to be viewed primarily as explanation, but as Love.

# III.
## *Who's Afraid of Giordano Bruno?*

### I.

Of the countless thousands of individuals across several centuries to be examined, tried and convicted by the Inquisition, the one man who is perhaps best remembered today for his fatal intersection with it is Giordano Bruno (1548–1600). It is no exaggeration to observe that almost every volume I have read in recent years concerning Mars, space exploration, or the search for extraterrestrial life has contained at least a passing reference to the man portrayed as the great 'martyr' of Science. In the assessment of one Bruno scholar:"The myth of Bruno grows daily." (de León-Jones 3) In the popular mythology, Bruno was persecuted for his Copernican views, burned at the stake for upholding an infinite number of worlds in an infinite universe. The sheer power of repetition has led many to accept this version of events, and thus Bruno has assumed a place within a broad mythology of the rise of science. In their recent book, *The Privileged Planet*, Gonzalez and Richards provide a concise summary of the broad outlines of this myth:

> Copernicus, according to the popular story, demoted us by showing that ours was a sun-centered universe, with Earth both rotating around its axis and revolving around the Sun like other planets. This claim is sometimes accompanied by still more egregious factual errors. For instance, Bruce Jakosky explains in *The Search for Life on Other Planets*,"Because of this tremendous change in world view, Copernicus' views

*This essay was originally presented to the 2005 convention of the Mars Society in Boulder, CO.*

were not embraced by the Church: the history of his persecution is well known." Never mind that Copernicus wasn't
persecuted and died the same year (1543) that his ideas
were published, not at the oil-soaked stake but peacefully
and of natural causes. Since these historical facts muddy the
popcorn-movie simplicity of the Official Story, with its cast
of intrepid, steely eyed scientific heroes on the one hand and
its one-dimensional villain priests on the other, the historical
facts are garbled. ...

      The popcorn movie continues on from Copernicus's
persecution with a bravura medley of fact and fiction: the
messiah Copernicus leaves his even less fortunate followers,
like Bruno, the first martyr, and Galileo, the first saint, to
suffer even more hideous consequences. In time, however,
the brave and unflagging march of scientific evidence over-
whelms the darkness and idiocy of religious superstition—
swelling and triumphant musical score followed by cheers
and the film's credits. (Gonzalez and Richards 222–3)

One of the most recent restatements of the myth can be
found in Dr. Ben Bova's work concerning the search for extraterrestrial life, *Faint Echoes, Distant Stars*. Declaring that Bruno "was rather
like the Carl Sagan of his era," Bova proceeded to observe,

      Above all else, Bruno broadcast the concept that
the universe was infinite in extent and that it must contain
untold number of worlds like our Earth—worlds people by
intelligent creatures. He was the first to make the point that
the stars must be other suns; the reason that they appear as
pinpoints of light is that they are immensely far away. ...
      The Church could not tolerate such a challenge. (14)

James Kargel (*Mars—A Warmer Wetter Planet*) repeats the same claim:
"When Galileo started his astronomical observations in 1609, the
ideas were already firmly rooted in Western science that the planets

are worlds, some like our own and some not, and that the vast real estate in outer space was not entirely barren. Bruno had been burn alive just years earlier for claiming that other worlds exist and that some may be habitable." (388)

Bruno entered the Dominican Order in 1563, but the troubles which came from his heretical views led him to flee the order in 1576. However, far from living 'on the run,' Bruno enjoyed the patronage of the French king, Henri III, and lectured in Geneva, Paris, Oxford, and Wittenberg, among other places before his arrest in 1592. Usually, Bruno left a city when he had worn out his welcome, hurling epithets as he went. Bruno was arrested and turned over to the Inquisition after he had settled in Venice, where he had been engaged as a private instructor. Although Bova insists that Bruno "refused to compromise or abandon his ideas," (15) the facts are quite different:

> At the end of the Venetian trial, Bruno fully recanted all the heresies of which he was accused and threw himself in penitence on the mercy of the judges. He had, however, by law to be sent to Rome while the case dragged on. In 1599, an effort was made to clarify the situation by the famous Jesuit, Robert Bellarmine, who, assisted by Tragaglio, drew up eight heretical propositions from his work which Bruno was required to abjure, and he said that he was prepared to do so. But later in the year he withdrew all his retractions, obstinately maintaining that he had never written or said anything heretical and that the ministers of the Holy Office had wrongly interpreted his views. ... He was burned alive on the Campo de Fiori in Rome on February 17, 1600. (Yates 349)

Although Bruno is now cast in the role of Copernican martyr, Bruno's 'Copernicanism' was not of a character which Copernicus' true intellectual descendants would want to profess today. In fact, it is more factual to say that Bruno *adapted* 'Copernicanism' to his predetermined views, rather than *adopting* true Copernicanism. Noted astronomer Owen Gingerich has conducted the only survey of all known copies of Copernicus' *De Revolutionibus*, and he offers the following insight concerning the supposed martyr for Copernicanism, Giordano Bruno: "Bruno had been sentenced as a heretic for a plethora of heterodox ideas, including the plurality of worlds, but he seemed at best rather ill informed about Copernicus' ideas. His *de revolutionibus* contained a bold signature but no evidence that he had actually read the book. In any event, his Copernicanism was not a major factor in his conviction." (65–66)

Current Bruno scholarship (e.g., Yates 1964, de León-Jones 1997, Webster 2005) offers a valuable corrective to the nineteenth century Bruno myth which is still so dominant in popular literature. The primary focus of Bruno's writings is Hermetic[1]/Egyptian magic and the study of the occult.

> Yet is has also been observed that "Bruno's worldview is vitalistic, magical; his planets are animated beings that move freely through space of their own accord like those of Plato or Patrizi. Bruno's is not a modern mind by any means." To this it may be added that ... Bruno's acceptance of Copernican earth movement was based on magical or vitalistic grounds, and that, not only the planets, but also the innumerable worlds of his infinite universe move through space like great animals, animated by the divine life. (Yates 244)

The occultism with which Bruno was obsessed is referred to as 'demonic' magic because it is focused on the notion that there are 'demons' or 'spirits' which animate the heavens, and which may be invoked or manipulated through the use of talismans and other magical arts or devices. Bruno considered himself quite adept in such magic, designing patterns for talismans for such magical purposes. Thus Bruno's beliefs and practices went far beyond the dreams of astrology and its practitioners; Bruno did not want to simply 'inter-

---

[1] Taking its name from the Gnostic writings of 'Hermes Trismegistus'.

pret' the heavens, he wanted to invoke the power of the demons he believed filled the heavens. Again, commenting on Bruno's De magia, Yates observes:

> Obviously a vital stage in this ascent would be to reach the demons, and Bruno's magic is quite frankly demonic. He knows the basis of the natural *spiritus* magic but entirely abandons Ficino's reservations. Bruno wants to reach the demons; it is essential for his magic to do so; nor are there any Christian angels within call in his scheme to keep them in check. Bruno, of course, like all magicians, regards his magic as good magic; only other people's magics are ever bad to the magician. And from his point of view and with his belief in the Egyptian religion as the right one, this would be the right magic, since, as we all know, it was the demons which the Egyptian priests knew how to manipulate and draw down into their statues. (264–5)

Far from Bruno being the 'Carl Sagan' of his generation, one might do better to find a comparison in the wilder fringes of the cults which are obsessed with extraterrestrials. The Copernican cosmology was not a 'scientific' model for Bruno; rather, it was an symbol waiting to be filled with mystical meaning.

Bruno's philosophy cannot be separated from his religion. It was his religion, the "religion of the world", which he saw in this expanded form of the infinite universe and the innumerable worlds as an expanded gnosis, a new revelation of the divinity from the "vestiges". Copernicanism was a symbol of the new revelation, which was to mean a return to the natural religion of the Egyptians, and its magic, within a framework which he so strangely supposed could be a Catholic framework.

Thus the legend that Bruno was prosecuted as a philosophical thinker, was burned for his daring views on innumerable worlds or on the movement of the earth, can no longer stand. That legend has already been undermined by the publication of the *Sommario*, which shows how little attention was paid to philosophical or scientific questions in the interrogations, and by the writings of Corsano and Firpo, laying stress on Bruno's religious mission. (Yates 355)

For Bruno, Copernicus was a "mere mathematician"; Bruno praised himself as the one who "has released the human spirit, and set knowledge at liberty." (Yates 236, 238) Thus all of Copernicus' mathematics were in need of a spiritual interpretation, and Bruno was eager to explain them in light of his 'Egyptian' Hermetico-magical religion: "Completely involved as he was in Hermetism, Bruno could not conceive of a philosophy of nature, of number, or geometry, of a diagram, without infusing into these divine meanings. He is thus the last person in the world to take as a representative of a philosophy divorced from divinity." (Yates 355–6)

Thus one is confronted with two 'Brunos.' There is, on the one hand, the Bruno of myth: prophet of Copernicanism and proto-scientist, a selfless idealist and model for future generations who bravely upheld the existence of an infinite number of worlds in an infinite universe and was martyred for that teaching. On the other hand, there is the Bruno of history: an extremely egotistical occultist, obsessed with magic and for whom the heavens are filled with powerful demons for the adept to manipulate. Either way, one readily acknowledges the obscenity of the Inquisition's judicial murder of opponents of the Roman Church. But when one recognizes why Rome burned Bruno—not on account of his superficial 'Copernicanism,' but because of his occultism—one can gain a better understanding of the Church's relationship with the concept of a 'plurality of worlds,' and the possibility of extraterrestrial life. When one examines the history of the belief that there could be more than one world, and that more than one world could be the abode of intelligent life, adherents of the Bruno myth will be in for a shock.

## II.

As noted above, Kargel recently observed that "When Galileo started his astronomical observations in 1609, the ideas were already firmly rooted in Western science that the planets were worlds, some like our own and some not, and that the vast real estate in outer space is not entirely barren. Bruno had been burnt alive just years earlier for claiming that other worlds exist and that some may be habitable." (388) But this assertion might lead the astute reader to ask the question: If such an idea was already "firmly rooted," why was Giordano Bruno—and Bruno alone—burned at the stake for such views? The answer is that he was not—that is, Bruno was not burned because he defended the notion of the plurality of worlds, but because of the occult views which he associated with that belief, and for a variety of heresies which had nothing to do with the Copernican hypothesis.

But the view that Christianity is somehow 'threatened' by the existence of many worlds, and especially by the existence of extraterrestrial life, is almost as firmly rooted in certain circles as adherence to the Bruno myth. Among the more balanced statements of this sort in found in Donald Goldsmith's *The Hunt for Life on Mars*:

> In its organized theology, the Judeo-Christian tradition has more difficulty accepting the notion of extraterrestrial life than most other major religions. ... Though we risk oversimplification, we may divide the Judeo-Christian response to extraterrestrial life into two basic categories, the Earth-centered and the greater-glory camps. The latter group sees life elsewhere in the universe as just another jewel in the crown of creation. For theologians in this category, the story of Genesis simply omits what may have happened on other worlds. In response to the news from Mars [regarding ALH 84001], the spokesman for the Catholic archbishops of France, Olivier de la Brosse, said, "Christian theology has

never supposed that only Earth and its inhabitants are the works of God. The Bible deals only with the history of humanity. If it is confirmed that life exists on Mars, for me this is not a supplementary proof but a manifestation of God's power. I conclude that God is still greater than I thought." (Goldsmith 234)

Often this view is pressed even more aggressively in the realm of science fiction. For example, in Gregory Benford's recent book, *The Sunborn*, one finds the following 'exchange':

She remembered the church her family had attended, pillars and vaulting white as plaster, like a cast around the broken bone of faith. Still . . . "Christianity has the most to lose from intelligent aliens, right? Jesus was *our* savior. Dolphins and gorillas and supersmart aliens—he didn't die for them."

"Um." Viktor sighed, resigned to a discussion. "Jesus was God's only son, yes?"

"The Bible says so."

"So unless God has the same son go around to every planet..."

"Or wherever these things we've found live—"

"Dying at every one of them? I am no expert, but—seems cruel."

"Worse, it means part of God has to go around dying all the time."

"Am glad I'm not a theologian." (234-5)

And so on. Centuries-old straw men are thus dispatched with arguments so timeworn that Benford even goes so far as to cite the eighteenth century pamphleteer Thomas Paine. One suspects that reliance on such dated argumentation is the result (at least in part) of a lack of familiarity with the history of the intersection of cosmology, theology and philosophy. However, the writings of Steven J. Dick and Pierre Duhem on the topic of the plurality of worlds are extremely

38

useful in meeting this need.[2]

If we are to understand the emergence of the idea of a plurality of worlds, then the crucial year for our consideration is not 1600 A.D. (the year when Bruno was burned) but 1277.

Beginning in the 12th century, the writings of Aristotle returned to prominence in Western thought. As Dick observes: "The De caelo itself became available to the Latin West about 1170 in a translation by Gerard of Cremona, the greatest of all translators from Arabic to Latin. Greek and Arab commentaries on De caelo, especially those of Simplicius (sixth century) and Averroes (1126–1198), were soon translated." It was on account of Aristotle's De caelo that the notion of the plurality of worlds rose to prominence in medieval scholastic thought.

Thomas Aquinas and other theologians sought to accommodate Aristotle's philosophy to the service of Christian theology, and had (arguably) experienced some measure of success. However, according to Aristotle, it was impossible for the "First Cause" to create more than one world. In Aristotle's system, everything—according to the notion of the four elements—comes to rest in its natural place; thus there could not be more than one world because it would result in a violent tension between matter which was forcibly being divided from that which makes up the Earth. Based on the Aristotelean system, Aquinas (and others) maintained it would be impossible for there to be more than one world. However, such a teaching directly contradicted divine omnipotence. Therefore, in 1277, Etienne Tempier, the bishop of Paris, issued a condemnation of 219 opinions derived from Aristotelean principles, including the belief "that the First Cause cannot make many worlds." Dick observes:

---

[2] Dick observes that "no detailed historical explication of the idea of the plurality of worlds as related to Western science and culture has yet appeared. ... Only the work of the French historian and philosopher of science Pierre Duhem has treated the question, for a limited period (the Middle Ages), from the point of view of its relation to the history of concurrent scientific developments." (5)

Although the importance of the condemnation of 1277 to medieval science is a subject of controversy, its effect on the idea of a plurality of worlds is unmistakable. The insistence that God could indeed create other worlds if he wished meant two things: either God could act supernaturally, that is, above and beyond the Aristotelian laws of Nature, to produce another world, or Aristotelian physics could be altered sufficiently to allow for the possible existence of other worlds within natural law. Both of these strategies or some combination thereof, were in fact to be utilized. (28)

The Church thus took a stand against the prevailing cosmology of the day and opened the way for the development of the concept of a plurality of worlds. In the words of Pierre Duhem, "In favor of the opinion that the existence of several worlds is possible, Christianity fashioned an argument based on the creative omnipotence of God, an argument that pagan antiquity could not have foreseen." (448) Thus the Church's stand against manifest error in the Aristotelean system opened the way for a new cosmology:

The last quarter of the thirteenth century witnessed a remarkable turnabout for the fate of other worlds, as evidenced by the lectures and written commentaries on Aristotle composed at Paris and Oxford, medieval Europe's two great intellectual centers. At Paris Godfrey of Fontaine, Henry of Ghent, and Richard of Middleton held that a plurality of worlds was not theologically impossible; at Oxford William of Ware, Jean of Bassols, and Thomas of Strasbourg made the same claim. ... They were also the first to hold that God could create other worlds, and that every possible world could determine the natural motions of its own bodies, whether those bodies were different in form from those of our world or the same. Moreover, they held that God could create matter for another world *ex nihilo*. (Dick 28–29)

This teaching was of tremendous significance for the development of a doctrine of extraterrestrial life because the notion of a plurality of worlds brought with it the assumption that such worlds *would* have life. In this regard, the German theologian (and cardinal) Nicholas of Cusa (1401–1464) was the most substantial figure. Cusa clearly taught that other worlds would be the home to living beings:

> Life, as it exists here on earth in the form of men, animals and plants, is to be found, let us suppose, in a higher form in the solar and stellar regions. Rather than think that so many stars and parts of the heavens are uninhabited and that this earth of ours alone is peopled—and that with beings perhaps of an inferior type—we will suppose that in every region there are inhabitants, differing in nature and rank and all owing their origin to God, who is the centre and circumference of all stellar regions. (Dick 41)

Again, Nicholas taught: "It may be conjectured that in the area of the sun there exist solar beings, bright and enlightened intellectual denizens, and by nature more spiritual than such as may inhabit the moon—who are possibly lunatics—whilst those on earth are more gross and material. It may be supposed that those solar intelligences are highly actualized and little in potency, while the earth-denizens are much in potency and little in act, and the moon-dwellers betwixt and between." (Dick 42) Consider carefully what Nicholas of Cusa said: it is to be expected that the heavens would be full of life—and that perhaps of a higher nature, and that such life would be of a nature well suited to the world in which it lives. In the so-called 'medieval' period, one could maintain that there was more than one world, and that the life on Earth was quite possibly inferior to the life found in the heavens, and remain a cardinal in the Roman Catholic Church. It should also be noted that Cardinal Cusa taught that the heavens were infinite, having no absolute center. Nicholas is thus the first person to assert that life was universally present throughout an infinite cosmos.

(Dick 42) Although the cardinal's position may be more bold than that of others of his age, it is clearly documented that teaching that there was the possibility of a plurality of worlds (and thus the possibility of widespread life throughout creation) was required by the teaching of the Roman Catholic Church during the medieval period. Far from the notion of a plurality of world being hostile to Christian theology, Dick observes "the question of life beyond the Earth was part of the natural evolution of the Scholastic tradition." (*Plurality* 42)

As noted previously, Benford cited an argument by Thomas Paine in his effort to show the 'problem' for Christianity posed by extraterrestrial life. In Benford's words: "'I looked up this stuff, and there's even a quotation about Christianity and extraterrestrials from Thomas Paine, the American revolutionary [sic]—over two centuries old! He said'—she glanced at her notepad, on their side table—'Let's see, *He who thinks he believes in both has thought but little of either*. Ouch!'" (235) But of course the Church had spoken to this precise point long before the eighteenth century scribbler. Among other places, an answer can be found in the writings of William of Vorilong, a contemporary of Nicholas of Cusa. William was the first to pose the question, "whether men exist in that world [that is, another planet], and whether they have sinned as Adam sinned." This is, of course, to raise the central question for the Christian: would such beings stand in need of redemption? "Vorilong himself answered that such creatures 'would not exist in sin and did not spring from Adam. But it is shown that they would exist from the virtue of God, transported to that world, as Enoch and Elias in the earthly paradise.' Moreover, 'As to the question whether Christ by dying on this earth could redeem the inhabitants of another world, I answer that he is able to do this even if the worlds are infinite, but it would not be fitting for Him to go into another world that he must die again.'" (Dick 43)

There is no doubt that the medieval scholastic doctrine of the plurality of worlds has gone in some interesting directions over the centuries. But the proposition that somehow such a Catholic teaching, rooted ultimately in the omnipotence of the holy Trinity, would propose some insurmountable problem for the Church is not reasonable. To borrow a formulation, someone who believe in the existence of extraterrestrial life, but denies the existence of an omnipotent Creator has thought but little of either.

# IV.
# Implications of the Discovery of Extraterrestrial Life for the Christian Worldview.

## Introduction

We read in the first chapter of Ecclesiastes: "That which has been is what will be, that which is done is what will be done, and there is nothing new under the sun. Is there anything of which it may be said, 'See, this is new'? It has already been in ancient times before us. There is no remembrance of former things, nor will there be any remembrance of things that are to come by those who will come after." (1:9–11 NKJV) Our generation's speculations concerning the existence of extraterrestrial life are not new to modern thought; they are not new to the Western mind. As Benjamin Wiker observed: "It's rather surprising, perhaps, to find out that the debate about whether there is extraterrestrial life stretches back just shy of two and a half millennia."

Nevertheless, the discussion has gained a new immediacy as evidence continues to accumulate for extensive surface water on Mars in the distant past. Certainly, there are many interested observers who believe that such discoveries increase the likelihood that life once existed (or may even now exist) on Mars, and they recognize that the discovery of evidence of extraterrestrial life could have profound theological implications. In an article in the September 2003 issue of *The Atlantic Monthly*, Dr. Paul Davies declared, "...the discovery of just a single bacterium somewhere beyond Earth would force us to revise

---

*This essay was originally presented at the 2004 convention of the Mars Society in Chicago, Illinois.*

our understanding of who we are and where we fit into the cosmic scheme of things, throwing us into a deep spiritual identity crisis that would be every bit as dramatic as the one Copernicus brought about in the early 1500s, when he asserted that Earth was not at the center of the universe."

However, as dramatic as such sweeping statements might be, is Davies' assessment truly accurate? As C. S. Lewis, a prominent lay Anglican theologian, observed in 1958:

> Each new discovery, even every new theory, is held at first to have the most wide-reaching theological and philosophical consequences. It is seized by unbelievers as the basis for a new attack on Christianity; it is often, and more embarrassingly, seized by injudicious believers as the basis for a new defence.
>
> But usually, when the popular hubbub has subsided and the novelty has been chewed over by real theologians, real scientists and real philosophers, both sides find themselves pretty much where they were before. So it was with Copernican astronomy, with Darwinism, with Biblical Criticism, with the new psychology. So, I cannot help expecting, it will be with the discovery of "life on other planets"—if that discovery is ever made. (84)

Although two thousand years of speculation concerning extraterrestrial life are not a substitute for such an actual discovery, such theorizing is still far from useless, since it models potential implications for various possibilities. This essay endeavors to weigh some of the potential theological implications of such a discovery for a Christian worldview. Although such an effort requires much simplification, given the broad spectrum of viewpoints within the global Christian community, it is necessary to start *somewhere*, and given the cultural context of much of humanity's efforts at space exploration, a Western Christian perspective offers much to commend itself for our consideration.

### *The Significance of the Discovery of Extraterrestrial Life*

Even within the relatively narrower context of Western Christendom, one does not encounter a monolithic response to the concept of extraterrestrial life. In his book, *The Hunt for Life on Mars*, Astronomer Donald Goldsmith observed:

> In its organized theology, the Judeo-Christian tradition has more difficulty accepting the notion of extraterrestrial life than most other major religions. ... Though we risk over-simplification, we may divide the Judeo-Christian response to extraterrestrial life into two basic categories, the Earth-centered and the greater-glory camps. The latter group sees life elsewhere in the universe as just another jewel in the crown of creation. For theologians in this category, the story of Genesis simply omits what may have happened on other worlds. In response to the news from Mars [regarding ALH 84001], the spokesman for the Catholic archbishops of France, Olivier de la Brosse, said, "Christian theology has never supposed that only Earth and its inhabitants are the works of God. The Bible deals only with the history of humanity. If it is confirmed that life exists on Mars, for me this is not a supplementary proof but a manifestation of God's power. I conclude that God is still greater than I thought." (Goldsmith 234)

Certainly there is a degree of merit in Goldsmith's assessment. Undoubtedly there are individuals who fit neatly into either "Earth-centered" or "greater-glory" "camps." There are some who would probably experience a crisis of faith if life were found elsewhere; others readily anticipate such a discovery and would find it confirms their belief in the work of God.

However, the two views—"Earth-centered" and "greater-glory"—are by no means mutually exclusive: one may certainly believe that God is glorified by the existence of extraterrestrial life, while still maintaining that because the Earth is the abode of man, it is the planet of primary concern to man, both theologically and philosophically. The central concern for the Church is man's redemption from the punishment for sin (those thoughts, words and actions which are transgressions of the Law of God). The center of the Christian faith is found in the life, death and resurrection of Christ Jesus who, having come "for us men and for our salvation" (as the Church confesses in the Nicene Creed), His death made atonement for the sins of the whole world.

The existence (or nonexistence) of extraterrestrial life is of no essential concern to the reaffirmation or negation of any article of the faith. Just as the Holy Scriptures neither preclude nor teach the existence of planets orbiting around other stars, so, too, the same texts do not address the issue of life existing on other worlds. Even the Roman Catholic Church, with its broader—and arguably more speculative—authority attributed to its teaching magisterium (centered in the authority of the pope), has apparently found no compelling need to make a formal pronouncement on the subject of the existence of extraterrestrial life.

If evidence of life is discovered on Mars (whether through meteorites, as was suspected to be the case several years ago with ALH 84001), or through human or robotic exploration of the red planet, the potential theological impact of such a discovery will hinge on whether such life is indigenous (a 'second Genesis') or whether it originated on Earth, and spread by panspermia. As Malcolm Walter observed: "It is necessary to be cautious and write 'finding a fossil bacterium on Mars could be sufficient to demonstrate that ... we are not alone', because one further step in logic is needed: we would have to demonstrate that any life found on Mars had an origin independent of that on Earth and did not, for instance, arrive there aboard a meteorite from here. If we have only fossils to work with, making that distinction is likely to prove extremely difficult." (Walter 147) In fact, it is arguably most reasonable to assume that any evidence of life found on Mars should be presumed to be life which originated on

Earth, unless and until proven otherwise through legitimate, scientific study. Concerns regarding forward contamination are nontrivial, and if there is valid scientific and ethical concern about contamination from robotic and human exploration of Mars, we must anticipate the probability that microbes have already arrived in advance of us.

For Christians, the theological impact of finding transplanted Earth life on Mars would be essentially nil. The existence of indigenous Martian microbes would be of interest, of course, and one might readily anticipate that such a discovery would signal the victory of the 'greater-glory' position, or the rise of that view in combination with the 'Earth-centered' view. In Psalm 8, David sets forth a model of a worthy Christian reaction to such a discovery:

> O LORD, our Lord,
> How excellent is Your name in all the earth,
> Who have set Your glory above the heavens! ...
> When I consider Your heavens, the work of Your fingers,
> The moon and the stars, which You have ordained,
> What is man that You are mindful of him,
> And the son of man that You visit him?
> For You have made him a little lower than the angels,
> And You have crowned him with glory and honor.
> (Psa. 8:1, 3–5 NKJV)

The proper Christian response to all the creative works of the holy Trinity is one of wonder and awe and humility. The psalmist acknowledges the vast expanse of the heavens, that glorious creation of the holy Trinity; what is mankind is comparison? The greater the heavens are known to be, the more we are aware of the glorious working of God, and the more conscious we are of the love and mercy of God which He has shown to humanity, in spite of our fallen condition. The existence, or absence, of widespread microbial life in the solar system (or throughout the universe) does not change that awareness.

However, there are those for whom the failure to find microbial life, or at least to find advanced forms of life, will prove a crisis to their belief in the so-called Copernican Principle (or, more accurately, 'Principle of Mediocrity'). In their book, *Rare Earth*, Peter Ward and Donald Brownlee endeavor to make a case for the relative abundance of microbial life, and virtual absence of more advanced forms of life, throughout the universe. Their thesis, compellingly argued, is a powerful lesson in how astoundingly rare the conditions for advanced life might be. Such scarcity has, at least for the authors, profound philosophical significance.

> If the Rare Earth Hypothesis is correct—that is, if microbial life is common but animal life is rare—there will be societal implications, or at least some small personal implications. What will be the effect if news comes back from the next Mars mission that there is life on Mars after all—microbial to be sure, but life. Or what if, after astronauts voyage repeatedly to other planets in our solar system, or even to the dozen nearest stars, we find nothing more advanced than a bacterium? What if, at least in this quadrant of the galaxy, we are quite alone, not just as the only intelligent organisms but also as the only animals? How much of our striving to travel into space is the hope of discovering—and perhaps talking to—other *animalia*? (Ward and Brownlee 278–9)

> The continued marginalization of Earth and its place in the Universe perhaps should be reassessed. We are not the center of the Universe, and we never will be. But we are not so ordinary as Western science has made us out to be for two millennia. [sic.] Our global inferiority complex may be unwarranted. What if Earth is extremely rare because of its animals (or, to put it another way, because of its animal habitability)?

> The possibility that animal life may be very rare in the Universe also heightens the tragedy of the current rate of extinction on our planet. ... And if animals are as rare in the Universe as we suspect, it puts species extinction in a

whole new light. Are we eliminating species not only from
our planet but also from a quadrant of the galaxy as a whole?
(Ward and Brownlee 283)

The understanding that life is rare and precious fits well with a biblical understanding that care for the Earth is a *stewardship* entrusted
to humanity. "Then the LORD God took the man and put him in
the garden of Eden to tend and keep it." (Gen. 2:15 NKJV) Man's
transgression of his stewardship which has resulted from his fall into
sin has brought suffering upon the creation. As St. Paul wrote to
the Church in Rome: "For the creation was subjected to futility, not
willingly, but because of Him who subjected it in hope; because the
creation itself also will be delivered from the bondage of corruption
into the glorious liberty of the children of God. For we know that
the whole creation groans and labors with birth pangs together until
now." (Rom. 8:20–22 NKJV) Whether or not humanity finds life of
Mars, the pressing need for repentance and renewal is driven home
by the way in which our species has abused its stewardship.

## The Significance of the Discovery of Extraterrestrial Intelligence

As remote as the possibility remains that humanity will find
life on Mars, it is even *more* remote that we would ever find evidence of
extraterrestrial intelligent life within the confines of our solar system,
even if such life exists elsewhere. Nevertheless, the theological questions raised by the existence of such life are even more interesting
that those considered previously, and so we will include a few brief
thoughts on the topic in this discussion.

On a certainly level, the Church does not have an inherent
problem with the concept of other created intelligences besides human beings. The Scriptures sets forth the existence of non-human,
created intelligences throughout the Old and New Testaments: God's
Word calls them angels and demons. Angels play a significant role
in the history of salvation as messengers and agents of God.

However, as Prof. John Haught of Georgetown University has observed, "Religious thinkers have long entertained the idea of the existence of extraterrestrial intelligent 'worlds,' and not always in 'heaven' (the angelic hosts), but also in 'the heavens' as well." (296) Haught contends that the discovery of extraterrestrial intelligence would actually be quite affirming to theism:

> Certainly, at the very least, an encounter with alternative intelligent worlds would be one more in a series of great occasions modern cosmology has provided for theology to enlarge its sense of God and divine creativity. But contact with extraterrestrial beings (ETs) would also provide an opportunity for theology, on its part, to display the unitive power of radical monotheism. ... Radical monotheism—with its belief that all things, all forms of life, all peoples and all worlds have a common origin and destiny (in a God who creates and encompasses all beings impartially)—is still the surest ground we have for embracing that which at first seem alien. (297)

And Haught adds (with words which are quite significant to members of the Mars Society!): "And by virtue of the omnipresence of the one God we too would have an extended home in all possible worlds to which we might eventually travel." (297)

C. S. Lewis, however, was more skeptical of man's potential for fruitful contact with alien intelligences; the novels of his "Space Trilogy" (*Out of the Silent Planet, Perelandra,* and *That Hideous Strength*) are all cautionary tales concerning humanity's place in the universe. In his 1958 essay, "Religion and Rocketry," Lewis addressed the thorny question of whether God's message of salvation could be interpreted as having applicability to any extraterrestrial intelligence. He observed,

> It may be that Redemption, starting with us, is to work from us and through us.
>
> This would no doubt give man a pivotal position. But such a position need not imply any superiority in us or

any favouritism in God. The general, deciding where to begin his attack, does not select the prettiest landscape or the most fertile field or the most attractive village. Christ was not born in a stable because a stable is, in itself, the most convenient or distinguished place for a maternity.

Only if we had some such function would a contact between us and such unknown races be other than a calamity. If indeed we were unfallen, it would be another matter. (88)

It was man's capacity for evil, the fruit of his fall into sin, that most troubled Lewis concerning the possibility of humanity's interaction with another intelligent race:

We know what our race does to strangers. Man destroys or enslaves every species he can. Civilized man murders, enslaves, cheats, and corrupts savage man. Even inanimate nature he turns into dust bowls and slag-heaps. There are individuals who don't. But they are not the sort who are likely to be our pioneers in space. Our ambassadors to new worlds will be the needy and greedy adventurer or the ruthless technical expert. They will do as their kind has always done. What that will be if they meet things weaker than themselves, the black man and the red man can tell. If they meet things stronger, they will be, very properly, destroyed. (89)

Lewis' point is fundamentally sound: Humanity must better respect all life as *creation*, not simply as 'happy accident,' before we will be ready for any such encounter to end in anything but disaster and sorrow. The experience of our species the past several centuries—the age of materialism and supposed Enlightenment—has been one of growing alienation, purposelessness, and devastation. It will simply not do to point to the material prosperity of a culture; what are such things to the hundreds of millions of human beings cut down by the fanatical adherents of materialistic ideologies?

Having noted such cautions, one should note that study and speculation will continue. As science continues to verify how vanishing rare—how seemingly impossible—the rise of living beings (let alone *intelligent* ones!) truly is, the purposes of science and theology will well-coincide.

In 1638, John Wilkins published his *The Discovery of a World in the Moone* in which he wrestled with the possibility of intelligent life on the moon—which was not as far-fetched an idea in his age as it is in ours. Wilkins (who would later become an Anglican bishop) was quite familiar with the writings of Tycho Brahe and Johannes Kepler, readily citing them, as he also cited the classical philosophers and the theologians of the early Church. Some of his observations near the conclusion of the book bespeak the wisdom of the writer:

> Being content for my owne part to have spoken so much of it, as may conduce to shew the opinion of others concerning the inhabitants of the Moone, I dare not my selfe affirme any thing of these Selenites, because I know not any ground whereon to build any probable opinion. But I think that future ages will discover more; and our posterity, perhaps, may invent some meanes for our better acquaintance with these inhabitants. **'Tis the method of providence not presently to shew us all, but to lead us a long from the knowledge of one thing to another.** ... So, perhaps, there may be some other meanes invented for a conveyance to the Moone, and though it may seeme a terrible and impossible thing ever to passe through the vaste spaces of the aire, yet no question there would bee some men who durst venture this as well as the other. True indeed, I cannot conceive any possible meanes for the like discovery of this conjecture, since there can bee no sailing to the Moone, unlesse that were true which the Poets doe but feigne, that shee made her bed in the Sea. We have not now any *Drake* or *Columbus* to undertake this voyage, or any *Daedalus* to invent a conveyance through the aire. However, I doubt not but that time

who is still the father of new truths, and hath revealed unto us many things which our Ancestours were ignorant of, will also manifest to our posterity, that which wee now desire, but cannot know. (p. 207–9, emphasis added)

Wilkins' trust in providence, and his prudence in limiting the extent of his speculation, would serve us well today. If we are not alone, then science and theology both instruct us how fortunate we are, and warn us against hubris. If we are alone, in the sense that there are no extraterrestrials (intelligent or otherwise), then science and theology serve us well to teach us how precious life truly is.

# V.

# Several Theological Considerations Concerning Plans for Terraforming Mars.

## Introduction

Discussions of terraforming Mars usually focus on the scientific and engineering challenges presented by such a project. Certainly such a focus is understandable, because the technical challenges are daunting, to say the least. A variety of theoretical models has been proposed and these models will no doubt continue to be revised as scientific knowledge continues to expand.

However, important ethical discussions need to take place in the context of these scientific studies. A vital lesson of the last several decades of human experience is that not everything which *can* be done, *should* be done. For example, the mapping of the human genome presents opportunities for providing medical means for alleviating suffering, but also makes possible the most systematic discrimination in history. The knowledge of the atom may provide power, medical benefit, and the preservation of food resources, or it may exterminate thousands. Scientific knowledge, unless it be guided by ethical principles, will lead to monstrous ends. Thus, the idea of reshaping worlds in the service of mankind is not only the greatest engineering project presently imaginable; it is also one of the greatest ethical challenges potentially confronting us.

Dr. Russell Kirk noted that "The great line of division in modern politics ... is not between totalitarians on the one hand and

---

*This essay was originally presented to the 2000 convention of the Mars Society in Toronto, Canada.*

55

liberals (or libertarians) on the other; rather, it lies between all those who believe in some sort of transcendent moral order, on one side, and on the other side all those who take this ephemeral existence of ours for the be-all and end-all—to be devoted chiefly to producing and consuming." (Kirk, 280) It is our underlying contention that a transcendent moral order does exist, and that man's interaction with nature must be in keeping with such an ethical standard. We will, therefore, examine the influence ethics—especially a self-sacrificial, 'kenotic' ethic—should have on any decisions concerning terraforming in the event that life does exist on Mars, especially in light of the apparent intelligent design of the cosmos for the support of life.

## More than 'Good Luck'—Growing Support for Intelligent Design

Insight into the ethics of terraforming is further illuminated by a growing body of evidence supporting the conclusion that the cosmos was designed to support life. Indeed, the weight of the evidence for design is so strong, that one author recently declared:

The question posed by intelligent design is not how we should do science and theology in light of the triumph of Enlightenment rationalism and scientific naturalism. The question rather is how we should do science and theology in light of the impending collapse of Enlightenment rationalism and scientific naturalism. These ideologies are on the way out. They are on the way out not because they are false (although they are that) or because they have been bested by postmodernity (they haven't) but because they are bankrupt. ... They lack the resources for making sense of an information age whose primary entity is information and whose only coherent account of information is design. (Dembskilutheran@ix.netcom.com 15)

A survey of the evidence for intelligent design is far beyond the scope of this essay, but a brief summary of the argument is germane. Advances in molecular biology and information science, in particular, have led a growing circle of scholars to conclude that the evidence against the universe and life developing by 'accident' is overwhelming; rather, the data points toward a Designer who has ordered all of nature for the existence and sustenance of life, including intelligent life. Michael Denton's recent book, *Nature's Destiny—How the Laws of Biology reveal Purpose in the Universe*, is possibly one of the most comprehensive analyses of the scientific evidence supporting intelligent design. Denton readily acknowledges the significance this conclusion poses for the present prevailing paradigm:

> The new picture that has emerged in twentieth-century astronomy presents a dramatic challenge to the presumption which has been prevalent within scientific circles during most of the past four centuries: that life is a peripheral and purely contingent phenomenon in the cosmic scheme. These advances in astronomy and physics have established what for Newton and generations of natural theologians was only an affirmation of belief: that there is indeed a deep and necessary connection between virtually every characteristic of the cosmic stage and the drama of life. It is ironic that those very features of the cosmos that were so troubling to the astronomers of the early seventeenth century—its vast size and the apparently infinite number of stars stretched out across its immensity—which inclined Kepler to wonder, "How can all things be for man's sake?" and which seemed to render the earth an irrelevant mote of dust in the cosmic scheme, have turned out to be absolutely critical and essential for our existence. (Denton 15)

In his conclusion, Denton summarizes the mountain of evidence as follows:

> We may not have final proof that the cosmos is *uniquely* fit for life as it exists on earth—because the possibility of alternative life cannot yet be entirely excluded—but there is

no doubt that science has clearly shown that the cosmos is *supremely* fit for life as it exists on earth. For as we have seen, the existence of life on earth depends on a very large number of astonishingly precise mutual adaptations in the physical and chemical properties of many of the key constituents of the cell: the fitness of water for carbon-based life, the mutual fitness of sunlight and life, the fitness of oxygen and oxidations as a source of energy for carbon-based life, the fitness of carbon dioxide for the excretion of the products of carbon oxidation, the fitness of bicarbonate as a buffer for biological systems, the fitness of the slow hydration of carbon dioxide, the fitness of the lipid bilayer as the boundary of the cell, the mutual fitness of DNA and proteins, and the perfect topological fit of the alpha helix of the protein with the large groove of the DNA. In nearly every case these constituents are the only available candidates for their biological roles, and each appears superbly tailored to that particular end.

...

In short, science has revealed *a vast chain of coincidences which lead inexorably to life* on earth—not just microbial life but all life on earth, including large, air-breathing organisms like ourselves—a chain of adaptations which leads from the dimensions of galaxies, through the physical conditions in the center of stars to the heat capacity of water and the atom-manipulating capacities of proteins, and on eventually to our own species and our ability to comprehend the world. From the inertial resistance we encounter when we move our hand, determined by the mass of the most distant stars, to the radioactive heat in the earth's interior which drives the great tectonic system, thus ensuring a continual replenishing of the vital elements of life—all nature, every facet of reality, is bound together into one mutual self-referential biocentric whole. (Denton, 381–382)

Indeed, even our ability to analyze substantively the nature of the creation around us is evidence of design. Denton writes: "However, there is another intriguing aspect to our success—the mutual fitness of the human mind and particularly its propensity for and love of mathematics and abstract thought and the deep structure of reality, which can be so beautifully represented in mathematical forms. In other words, the logic of our minds and the logic of the cosmos would appear to correspond in a profound way. And it is only because of this unique correspondence that it is possible for us to comprehend the world." (Denton 259)

Whereas Denton points to the astounding number of 'co-incidences' necessary for the rise of life, Michael Behe points to the existence of irreducibly complex biomolecular systems as sufficient evidence of intelligent design.

> With the advent of modern biochemistry we are now able to look at the rock-bottom level of life. We can now make an informed evaluation of whether the putative small steps required to produce large evolutionary changes can ever get small enough. ... the canyons separating every-day life forms from their counterparts in the canyons that separate biological systems on a microscopic scale. Like a fractal pattern in mathematics, where a motif is repeated even as you look at smaller and smaller scales, unbridgeable chasms occur even at the tiniest level of life. (Behe, 15)

Complex systems, because they require numerous highly specified elements to function, are essentially impossible to explain in evolutionary terms, Behe declares, because they are utterly non-functional unless all elements are originated simultaneously; thus gradual, unguided development through "survival of the fittest" cannot develop such mechanisms. As Behe observes,

> By *irreducibly complex* I mean a single system com-posed of several well-matched, interacting parts that contrib-ute to the basic function, wherein the removal of any one of

the parts causes the system to effectively cease functioning. An irreducibly complex system cannot be produced directly (that is, by continuously improving the initial function, which continues to work by the same mechanism) by slight, successive modifications of a precursor system, because any precursor to an irreducibly complex system that is missing a part is by definition nonfunctional. ... Since natural selection can only choose systems that are already working, then if a biological system cannot be produced gradually it would have to arise as an integrated unit, in one fell swoop, for natural selection to have anything to act on. (Behe 39)

Nature is filled with many examples of irreducibly complex biological systems and each such system, Behe argues, stands as a witness against unguided, 'natural' selection. Behe analyzes the specific details of several such systems, including the bacterial flagellum, the coagulation of blood, and the immune system. The complexity of such systems leads Behe to conclude:

To a person who does not feel obliged to restrict his search to unintelligent causes, the straightforward conclusion is that many biochemical systems were designed. They were designed not by the laws of nature, not by chance and necessity; rather, they were *planned*. The designer knew what the systems would look like when they were completed, then took steps to bring the systems about. Life on earth at its most fundamental level, in its most critical components, is the product of intelligent activity.

The conclusion of intelligent design flows naturally from the data itself—not from sacred books or sectarian beliefs. Inferring that biochemical systems were designed by an intelligent agent is a humdrum process that requires no new principles of logic or science. It comes simply from the hard work that biochemistry has done over the past forty years, combined with consideration of the way in which we reach conclusions of design every day. (Behe 193)

William Dembski builds on Behe's arguments, expanding it to the broader problem of explaining the existence of complex-specified information in terms of naturalism. The question which must be answered is:

> Where does CSI [complex-specified information] come from, and where is it incapable of coming from? ... Algorithms and natural laws are in principle incapable of explaining the origin of information. To be sure, algorithms and natural laws can explain the flow of information. Indeed, algorithms and natural laws are ideally suited for transmitting already existing information. What they cannot do, however, is originate information. ... Functional relationships at best preserve what information is already there, or else degrade it—they never add to it. (Dembski 160–161)

The only answer explaining the origin of CSI, according to Dembski, is intelligent design. But if the cosmos is designed, it raises the possibility of ethical constraints on our actions which are quite different from those presented by a doctrine of "survival of the fittest." As Dembski explains,

> Design also implies constraints. An object that is designed functions within certain design constraints. Transgress those constraints and the object functions poorly or breaks. Moreover we can discover those constraints empirically by seeing what does and doesn't work. This simple insight has tremendous implications not just for science but also for ethics. If humans are in fact designed, then we can expect psychosocial constraints to be hardwired into us. Transgress those constraints and we personally as well as our society will suffer. There's plenty of empirical evidence to suggest that many of the attitudes and behaviors our society promotes undermine human flourishing. Design promises

to reinvigorate that ethical stream running from Aristotle through Aquinas known as natural law. (Dembski 151)

Such ethical constraints, implicit in a world which exists because of intelligent design, should strongly influence decisions made regarding terraforming in the event that life is discovered on Mars. If life exists as part of a plan with a purpose, then the discovery of life on Mars, and our response to that discovery, will pose one of the most profound ethical challenges in history.

## *Possibilities for Life on Mars*

Of all of the potential ethical factors involved in the decision to terraform Mars, the most important factor would be indigenous life and the impact (if any) which terraforming could have on that life. The on-going debate over the ALH84001 meteorite is sufficient evidence that the existence of life on Mars in the past or present is a controversy which is far from settled, and constitutes sufficient scientific grounds to justify manned expeditions to the red planet. Hopefully, such explorations will yield data which permits general scientific consensus on one of four potential alternatives: (1) there has never been life on Mars; (2) life existed on Mars in the past, but has since died out; (3) life presently exists on Mars through panspermia from Earth; or (4) life truly 'native' to Mars presently exists.

In the case of the first alternative—the absence of life at every phase of Martian history—does not pose any ethical issue *vis-à-vis* Martian life, unless one wishes to debate the ethics of introducing life into a presently lifeless environment. (This is not to say that no other ethical issues could potentially arise, just that indigenous life is clearly eliminated as an issue.) Again, the second alternative—extinct Martian life—obviously poses no ethical questions as pertains to the impact terraforming would have on such life. For purposes of our discussion, the relevant alternatives are those which consider extant life (whether native or introduced by panspermia), and the effect terraforming would have on its future. Presumably few would raise ethical objections if terraforming benefited both humanity and Martian life, but the question arises: If life is discovered on Mars, and it was determined that terraforming would have a negative effect on the continued existence of such life, could humanity still proceed to terraform Mars?

## Some Implications for Intelligent Design on the Decision to Terraform Mars

In *The Search for Life on Mars*, Malcolm Walter declares that 'forward contamination' (the introduction of Earth life into the Martian environment) "can be treated as an ethical issue analogous to the protection of wilderness areas on Earth." (Walter 137) If this is indeed the case, then the ethical issue of Martian life is even more significant. In the event that life has been introduced to Mars from Earth through panspermia, then Walter's analogy holds, and such Earth life will be worthy of study and preservation for the same reasons that we protect rare life on Earth. However, if Martian life is not related to Earth life—if such life is truly alien to us—then the challenge is even more profound.

If Martian life is related to Earth life, science will be well-served by studying such life in a way which looks for its *purpose*. Just as scientists search rain forests looking for potential treatments for human diseases (a somewhat odd methodology unless one believes that nature is designed), astrobiologists should study such transplanted Earth life toward the end of learning that which might aid our own survival. As regards terraforming, however, it is unlikely that such climactic changes would be detrimental to panspermic Earth life; rather, it would probably greatly benefit from our efforts.

Nancey Murphy and George F. R. Ellis set forth a kenotic ethic in their book, *On the Moral Nature of the Universe*, which may prove helpful in the event that Martian life is not 'imported.' The core of their kenotic ethic is: "Self-renunciation for the sake of the other is humankind's highest good." (118) Although one may not agree with all of their corollaries to the core thesis, one observation of their ethic is particularly interesting in connection with our topic: "The final level in this progression is when I am willing to sacrifice goods or other aspects of my own welfare for others who are not my own kin and whose employment will not contribute directly to my own happiness (indeed, I may not even know them). We see this, for example, among those in various environmental movements who sacrifice some measure of their own standard of living for the sake of the ecosystem and for future generations." (127)

It is hard to imagine a situation where terraforming efforts could harm indigenous Martian life; almost certainly anything capable of enduring the present hostile environment would either endure or thrive in the conditions which would exist because of terraforming. But it is hypothetically possible that terraforming may confront us with a moral dilemma of choosing between what we believe to be in our own best interests, and the continued existence of a truly alien life form. In such a situation, the ethical maturity of our species would gain through such a show of mercy would outstrip the economic benefits. In such a circumstance we are, at most, stewards—not masters—over another's creation. As Murphy and Ellis declare:

> We argue that in light of a theological account of ultimate reality, which includes God's *moral* purposes for the universe, the anthropic features of the universe (i.e., features tending toward the appearance of humans) can now be interpreted as the necessary conditions not only for life but for intelligence and freedom. The anthropic universe is seen to be, more precisely, a moral universe. Once the universe is seen to be a moral universe, it becomes possible to explain added cosmological features that other (nontheistic) accounts of the anthropic features cannot explain: Why is there a universe at all? and, Why is it lawlike? (203)

The essence of ethical action is doing the 'right thing' precisely when it is difficult. Life is not a gift which we can create; it is a tremendous blessing which is to be treasured and preserved. May our humility be that of the Psalmist:

> When I consider Your heavens, the work of Your fingers,
> > The moon and the stars, which You have ordained,
> What is man that You are mindful of him,
> > And the son of man that You visit [care for] him?
> For You have made him a little lower than the angels,
> > And You have crowned him with glory and honor.
> > > (Psalm 8:3–5)

# VI.

# "Be Fruitful and Multiply"— Divine Creation and the Motivation to Colonize Mars.

## I. The Need for Christian Involvement.

The recent thirtieth anniversary of the Apollo 11 mission provided a wonderful opportunity to reflect on the exercise of public commitment which was needed to reach the Moon. Of course, the program did not enjoy universal support, but it drew support for many different reasons. However, as the attention of the media (and, presumably, a significant portion of the public) found their initial fascination wearing off, interest in the later Moon landings began to wane.

There is an obvious lesson for us in the fickle reaction to the Apollo program. Given the resources which will be required to explore and colonize Mars, it will be necessary to explain the need for such endeavors in terms which will be acceptable to the broadest feasible segments of society. It is one thing to generate the degree of support needed for what some refer to as a "flags and footprints" mission; it is quite a different matter to keep up *and increase* the level of spiritual commitment needed to commit to, and accomplish, the creation of a Mars colony.

The choice of describing our difficulty as one of "spiritual commitment" was not accidental. The Founding Declaration of the

*This essay was originally presented to the 1999 convention of the Mars Society in Boulder, Colorado.*

Mars Society declares, "Civilizations, like people, thrive on challenge and decay without it. ... As the world moves toward unity, we must join together, not in mutual passivity, but in common enterprise, facing outward to embrace a greater and *nobler* challenge than that which we previously posed to each other." Again: "The settling of the Martian New World is an opportunity for a *noble* experiment in which humanity has another chance to shed old baggage and begin the world anew; carrying forward as much of the best of our heritage as possible and leaving the worst behind." (emphasis added) To speak of something as "noble" is to proclaim that it shows high moral qualities, or a greatness of character. Nobility is a matter of the spirit. Yes, we may speak of noble metals and noble gases (both of which will be of great value to any Mars colony), but nobility of character—nobility of spirit—is of greater value to any society than platinum. Contrary to the views of some, mind and spirit are not some accidental quirks of evolutionary development, nor is the character of our spirit merely the expression of biological needs (Darwin) or economic forces (Marx). It is often the occasions of spirit denying the needs or desires of the flesh which we call virtues. There is more than a *quantitative* difference between a swarming, dividing hive of bees and the act of colonizing another world; the difference is *qualitative* as well. In the one situation, there is no choice—instinct, not choice or conscience, determines action. In the other situation, the decision is made to act in a way deemed to be noble or moral, despite the personal cost. If nothing more than a supposed biological mandate for perpetuation of the species drives our efforts, then one could no more declare it to be "noble" than to call breathing and eating "noble"—such acts are necessary, not noble.

Science is important—but we are not proposing to go to Mars only for scientific knowledge. Prosperity can be a great blessing—but we are not proposing a Martian colony simply for the acquisition of wealth. We are not going as a demonstration of national might. And let us hope that we will not focus now and in the years to come simply on the questions of "how"—questions of technology and finances—but that we will also direct ourselves even more to questions of "why"—why we should go, or not go; why we should colonize or not colonize; why we should terraform or not terraform, and so on. An action is neither "good" nor "noble" simply because we are capable of the action—certainly the twentieth century should have taught mankind this lesson. Just as man is both body and spirit, and neglecting the needs of either parts damages the whole, so neglecting questions either of "how" or "why" will damage what we propose to accomplish.

A year ago [1998] the author was privileged with the opportunity of addressing the society on the topic, "A Shining City on a Higher Hill: Lessons from the Last Colonization of a 'New World'." Working from the model of the early seventeenth century colonization of North America, I endeavored to weigh the importance of three motivations to colonization: (1) military expansion or competition between colonizing nations, (2) economic growth or exploitation, and (3) pursuit of political and religious freedom. The precise role of Christianity in the various aspects of the formation and development of the American culture will almost certainly continue to be a matter of great debate long after we have all become dust. The matter is certainly complicated because the religion has exerted influence beyond its overt manifestations, to shape philosophy, to filter the influence of the classical age, indeed to shape language itself. Nevertheless, as I demonstrated a year ago, neither competition between colonizing nations nor economic exploitation proved as definitive to the opening of the "New World" as the emigration of the Pilgrims and Puritans from England. In the words of the historian Paul Johnson, it was "the single most important formative event in early American history, which would ultimately have an important bearing on the crisis of the American Republic." (28) As the present writer wrote in a recent article,

> The Pilgrims (and the Puritans who followed) were, therefore, different from those who came before: They did not come as individuals, but as a community. They did not

come as adventurers, but as "planters" (colonists). They came with a specific *vision* motivating their settlement—a revitalization of their Christian faith—and understood themselves bound up in a covenant with God in this task. (22)

A legacy of the foundational role of Christianity in American society is the continued high level of Church affiliation. Public opinion polls regularly find that 95 percent of Americans believe in God and four in five describe themselves as "Christian."[*] Even those describing themselves as "non-Christian" often subscribe to uniquely-Christian teachings, such as the Resurrection (52%) and the virgin birth (49%).[†] Indeed, over 120 million Americans are members of the ten largest denominations in the United States, and approximately 246,319,000 of all North Americans are Christians. (Manske and Harmelink 34 and C) Globally, over 1.9 billion people are members of various Christian denominations, accounting for 33.6% of the world's population. (Manske and Harmelink B)

Given the role believing communities played in opening the last "New World," and given the number and influence of such believers today, it seems logical that an effort should be made to address their potential motivations for supporting Mars exploration and colonization. This essay is less concerned with motivations common to all people (such as a desire for economic prosperity); rather, I will endeavor to examine specifically Christian motivations for Mars colonization, especially those related to the doctrine of creation.

## II. The Motivation for Christian Involvement.

Aside from some of the common, secular "positive" motivations (motivations which would draw men to Mars) such as potential economic gain, there are also "negative" motivations having to do less

---

[*] "Poll finds Christians questioning religion's principles," The Indianapolis Star, 12 September 1994, p. A5.
[†] ibid.

with getting into space than they have to do with getting off Earth. Most prominent among these motivations is what could be called the "don't have all your eggs in one basket" theory. This theory postulates a global threat—a comet, massive asteroid, or uncontrollable plague—which threatens all of mankind on earth. It makes sense not to "keep all our eggs in one basket," the argument goes, "so we'll establish a Mars colony, Moon base, etc. to guarantee that mankind goes on, no matter what happens to Earth."

The supposed "threat" of doomsday scenarios is a motivation unlikely to be particularly effective with traditional Christians, whether Roman Catholic, Eastern Orthodox, Lutheran, or conservative (especially Evangelical) Protestant if one is seeking to justify space colonization. Christian eschatology* doesn't find, frankly, much to fear in such scenarios. Christians have always lived in the conviction that the world will end suddenly at Jesus' return. As St. Peter wrote in his Second Epistle: "But the day of the Lord will come as a thief in the night, in which the heavens will pass away with a great noise, and the elements will melt with fervent heat; both the earth and the works that are in it will be burned up. Therefore, since all these things will be dissolved, what manner of person ought you to be in holy conduct and godliness, looking for and hastening the coming of the day of God, because of which the heavens will be dissolved, being on fire, and the elements will melt with fervent heat? Nevertheless we, according to His promise, look for new heavens and a new earth in which righteousness dwells." (3:10-13 NKJV) That some outside the Church deprecate this teaching, pointing out that nearly 2,000 years have passed without Christ's return, doesn't shake this eschatological expectation, for St. Peter also declared, "scoffers will come in the last days, walking according to their own lusts, and saying, 'Where is the promise of His coming? For since the fathers fell asleep, all things continue as they were from the beginning.'" (2

---

* The articles of doctrine concerned with "last things," i.e., the Second Coming, the Judgment, eternal death and eternal life.

Ptr. 3:3–4) This eschatological expectation is far from a depressing aspect of Christianity; rather, belief in the resurrection is a central hope of the faith. Christians, in the words of the Nicene Creed, "look for the resurrection of the dead, and the life of the world to come." As St. Paul wrote, "And if Christ is not risen, then our preaching is empty and your faith is also empty. ... But now Christ is risen from the dead, and has become the firstfruits of those who have fallen asleep." (1 Cor. 15:14, 20) Thus, trying to alarm Christians with threats of the death of mankind have very limited appeal; we knew the clock was ticking down long before anyone realized there were rocks out there just waiting to "do us in."

More probable motivations are those which might be categorized as "negative" motivations (those which might push Christians away from Earth) or "positive" motivations (those which might draw them toward colonization) for Christian involvement in Mars colonization. One "negative" motivation is that of flight from oppression, one of the most powerful motivations in the seventeenth century colonies. Granted that religious freedom was a significant motivation for the successful colonization of the New World, is it possible this motivation would be operative in a future colonization of Mars? The stark reality is that Western Christians are probably experiencing a greater sense of cultural isolation and alienation than their seventeenth century forefathers. As John Lewis observes in Mining the Sky, "It was the search for freedom of religion that brought most of our ancestors here, and it will be the search for freedom from religious, political, and ethnic persecution that will send the first colonists forth into space." (240)*

If one is to engage positively Christians with the aim of gaining their support of Mars exploration and colonization, one must speak in terms of the doctrine of creation; that is, the belief that "In the beginning God created the heavens and the earth." (Gen.

---

*For a more comprehensive examination of "negative" or "flight" motivations, readers are referred to the eighth essay, and to the author's article in the November/December 1998 issue of *Ad Astra* magazine.

1:1) Again, faith in the Triune God as Creator is fundamental to all Christians, as finds expression in the Nicene Creed, "I believe in one God, the Father Almighty, Maker of heaven and earth and of all things visible and invisible." Christians universally hold this to be a doctrine rooted in the Holy Scriptures. It is important to note that we are not discussing the contended matter of the manner in which God created the cosmos; that is, whether Genesis 1 and 2 are to be read figuratively or literally. Christians have differed on this question and will probably continue to do so. The issue here is not so much an argument between 'young-earth'* versus 'old-earth'† Creationism. Rather, what is needed is an appeal to Christians rooted in a two-fold understanding of creation as (1) giving testimony to its Creator, and (2) God placing a requirement on mankind for careful stewardship of the creation.

The Church has long drawn comfort from the creation giving testimony to its Creator. Creation gives testimony both to the infinite power of the Creator and His love for His creation, especially those whom He created in His own image. As King David wrote in the eighth Psalm, "When I consider Your heavens, the work of Your fingers, the moon and the stars, which You have ordained, what is man that You are mindful of him, and the son of man that You visit him? For You have made him a little lower than the angels, and You have crowned him with glory and honor." (v. 3–5) Again, we hear in Psalm 19, "The heavens declare the glory of God; and the firmament shows His handiwork. Day unto day utters speech, and night unto night reveals knowledge. There is no speech nor language where their voice is not heard. Their line has gone out through all the earth, and their words to the end of the world." (v. 1–4) This belief that God reveals Himself in the creation is a powerful motivation for Christians to study that good creation. It is the pagan Gnostics—whether of the first century or the twentieth—who show contempt for the

---

* The contention that the earth is only a few thousand years old.
† The contention that the earth is billions of years old.

creation. The Christian knows that he beholds the good handiwork of God when he studies nature. In the words of John Calvin, "When a person, from beholding and contemplating the heavens, has been brought to acknowledge God, he will learn also to reflect upon and admire his wisdom and power displayed on the face of the earth, not only in general, but even in the minutest plants." (Zachmann 308) As the seventeenth century astronomer (and Lutheran) Johannes Kepler declared:

> I implore my reader not to forget the divine goodness conferred on mankind, and which the psalmist urges him especially to consider. When he has returned from church and entered on the study of astronomy, may he praise and glorify the wisdom and greatness of the Creator. ... Let him not only extol the bounty of God in the preservation of living creatures of all kinds by the strength and stability of the earth, but also let him acknowledge the wisdom of the Creator in its motion, so abstruse, so admirable.
>
> (quoted in Gingrich 28)

Or in the words of Dr. Herbert Uhlig, Professor Emeritus in the Department of Materials Science and Engineering, Massachusetts Institute of Technology, "Faith in the concept of a God who is concerned with His creation is essential to human hope, an optimistic world view, and ultimate survival of the human race. Any contrary view aligns humanity with the frustration of a drifting, meaningless universe facing a despondent future." (quote in *Cosmos, Bios, Theos* 126)

Just as importantly, the Christian believes mankind is responsible for exercising a divinely-mandated stewardship over the creation. As we read in Genesis 1, "Then God blessed them, and God said to them, 'Be fruitful and multiply; fill the earth and subdue it; have dominion over the fish of the sea, over the birds of the air, and over every living thing that moves on the earth.'" (Gen. 1:28) Even after the fall into sin, this stewardship continues, but in a more troubled state: "Cursed is the ground for your sake; in toil you shall eat of it all the days of your life. Both thorns and thistles it shall bring forth for you, and you shall eat the herb of the field. In the sweat of your face you shall eat bread till you return to the ground, for out of it you were taken; for dust you are, and to dust you shall return." (Gen. 3:17b–19) The creation is understood to be a good gift of God, which must be cared for, nurtured, and, because of man's fall into sin, restored. Misuse of the creation is not simply imprudent; it is wicked. Again, to quote Calvin,

> Let him who possesses a field, so partake of its yearly fruits, that he may not suffer the ground to be injured by his negligence; but let him endeavor to hand it down to posterity as he received it, or even better cultivated. Let him so feed on its fruits, that he neither dissipates it by luxury, nor permits it to be marred or ruined by neglect. Moreover, that this economy and this diligence, with respect to those good things which God has given us to enjoy, may flourish among us; let everyone regard himself as the steward of God in all things which he possesses. Then he will neither conduct himself dissolutely, nor corrupt by abuse those things which God requires to be preserved. (quoted in Zachmann 311–312)

In all man's stewardship endeavors, however, the Christian lives with the understanding that the completion of the restoration must await the fulfillment of the ages. As St. Paul writes in Romans 8, "For the earnest expectation of the creation eagerly waits for the revealing of the sons of God. For the creation was subjected to futility, not willingly, but because of Him who subjected it in hope; because the creation itself also will be delivered from the bondage of corruption into the glorious liberty of the children of God. For we know that the whole creation groans and labors with birth pangs together until now. Not only that, but we also who have the firstfruits of the Spirit, even we ourselves groan within ourselves, eagerly waiting for the adoption, the redemption of our body." (v. 19–23)

The zeal both for the study of nature, and for its preservation and cultivation, are noble traits which Christians would bring to Mars colonization (and, I suspect, will lead to a "Red" versus "Green" debate —my apologies to Kim Stanley Robinson—all our own). What can be learned to the glory of the Creator, and the work which can be done by way of careful stewardship of His creation, can be powerful motivations for Christians. In the words of Dr. Schawlow (recipient of the Nobel Prize for Physics and a professor of Physics at Stanford University),

Science cannot either prove or disprove religion. Religion is founded on faith. It seems to me that when confronted with the marvels of life and the universe, one must ask why and not just how. The only possible answers are religious. For me that means Protestant Christianity, to which I was introduced as a child and which has withstood the tests of a lifetime.

But the context of religion is a great background for doing science. In the words of Psalm 19, "The heavens declare the glory of God and the firmament showeth his handiwork". Thus scientific research is a worshipful act, in that it reveals more of the wonders of God's creation. (quoted in *Cosmos, Bios, Theos* 105–106)

## Conclusion

We have sketched in broad outline some of the positive motivations for Christians to take an active interest in Mars exploration and colonization. Much more needs to be said, of course, and this paper is only intended to be a brief introduction to this entire topic.

When seeking to reach out to Christians to solicit their support for Mars colonization, arguments should take their world view into account. Appeals should center in (1) learning what we can of the Creator through the creation, as well as (2) the cultivation of creation and its benefits to mankind on Earth. Such appeals should be based in the Scriptures, the Church fathers, and later significant theologians—preferably theologians representing as broad a historical perspective as possible. Representative members of the various denominations should enter into dialog with each other and members of their own church bodies to assess what efforts can be made for outreach. Toward this end, the Mars Society should focus on involving representatives of various denominations to present such appeals to members of their own communities.

As I noted at the beginning of this paper, the challenge before us is one of spiritual commitment. Certainly the world does not seem to lack either the resources or the technological prowess to accomplish both the exploration and colonization of Mars, God-willing. The prospect of colonizing another planet is arguably a noble one, but it is also one which should evoke another virtue—humility; humility that it is granted to us to live in this generation, and humility in order that hubris not pervert or doom our efforts. If the years ahead of us do not lead us to seek answers to "why," and not just "how"; if we do not seek to know better both ourselves and our Creator, then we will have missed a rare opportunity for growth in spirit. In this labor, as in all others, our cry is, "Soli Deo gloria!" As Kepler prayed:

> If I have been allured into brashness by the wonderful beauty of Thy works, or if I have loved my own glory among men, while advancing in work destined for Thy glory, gently and mercifully pardon me: and finally, deign graciously to cause that these demonstrations may lead to Thy glory and to the salvation of souls, and nowhere be an obstacle to that. Amen. (quoted in Gingrich 32)

# VII.

# A Theological Perspective on the Exploration and Colonization of Mars.

## I. The Universality of Metaphysical Motivations in Human Life.

From the founding convention in 1998, I have been greatly impressed by the rich variety of knowledge displayed by the presenters. Over the course of years, the Mars Society has gathered a talented and highly motivated membership which is eager to contribute to the vision which drives the society. The questions which drive the presenters, whether they are economists, artists, philosophers, educators, biologists, physicists, or computer engineers are: "How may we explore and settle Mars?" and "How can I best contribute to that effort?"

However, it has become increasingly clear as the years have progressed that the exploration and colonization of Mars is not simply a matter of solving the technical problems posed by such an endeavor, daunting though they may be. Exploring Mars is not only a matter of raising enough money to fund the research and development. Driving the agenda forward is a far more difficult task than lobbying representatives in various governments and giving presentations at the local library. Please understand me: all of these tasks are very important, and without them it is hard to imagine how the exploration and colonization of the Red Planet could possibly proceed. But the point I wish to highlight today is that for every one

_This essay was originally presented at the 2003 convention of the Mars Society in Eugene, Oregon._

of us, there are underlying motivations which guide us, compelling our interest, motivating us to travel to a convention such as this, and fueling the dream of a new frontier.

Becoming an advocate for the exploration of Mars is the outcome of preceding values and judgments. A goal such as opening a frontier on a new world is an application of a principle which precedes the goal. The moral purpose which precedes the goal is what gives meaning to our lives, and such purpose, if it is worthy of the title, must transcend us. Sociologist Douglas Porpora declares, "The fact is that moral purpose is less something we choose than something that chooses us. Before we ever choose to devote our lives to one or another moral purpose, that purpose must first move and inspire us." (8–9)

No matter whether one is an atheist, pantheist, or theist, an individual's world view is shaped by metaphysical foundations which are connected to a sense of purpose which leads us toward our goals. Even a person of the most rationalistic viewpoint must rely on non-verifiable epistemic principles (Schäfer 4) to formulate his understanding of the world. But as Huston Smith recently wrote: "With us, life's problems press so heavily on us that we seldom take time to reflect on the way our unconscious attitudes and assumptions about the nature of things affect the way we perceive what is directly before us." (Smith 25)

For the atheist or agnostic, the compelling motivation for exploring and inhabiting the Red Planet is certainly motivated by an interest in expanding mankind's knowledge and is also often rooted in a desire for the survival of the human species. One reads a great deal concerning the dangers of environmental degradation and even more exotic dangers to the planet such as comets. Although such an individual may deny humanity lacks corporate or individual teleological significance, nevertheless he or she supports the interplanetary expansion of the human race either as a moral good, or simply as an inevitability if the species is to thrive.

For the pantheist, viewing all of creation as being divine in some sense, there may be a desire to learn more about the "All" through knowledge, and even habitation, of Mars. Although some individuals within the Mars Society of the more pantheistic viewpoint subscribe to various conceptions of the Gaia thesis, viewing (to varying degrees) the earth as substantially a single organism, they believe the expansion of Earth's life to a (presumably) dead world is a positive development, increasing the portion of the universe which is 'self-aware' (i.e., through humanity). Generally the pantheistic worldview has not been as prevalent within the Mars Society as either the atheistic/agnostic or theistic perspectives are, and so it is more difficult to speak to their specific motivations based on direct conversation.

For the theist, and for the Christian in particular, the motivations are often quite different. Although there are individuals within the Church who may respond to either survival/expansion motivations, or to the virtue of expanding life throughout the cosmos, several other virtues play a role which differs from the motivations of others or which assume a particular character within the Christian life. It was the desire of Christian humanists to observe the hand of God within the creation which inaugurated and sustained the development of the sciences; belief that the creation is the work of God, and that man is responsible for his stewardship of that good creation, are central articles of the faith, reflected in the first article of the Apostles' Creed: "I believe in God the Father Almighty, Maker of heaven and earth." Exploration of Mars then occurs for the Christian in light of the knowledge he will gain, not only toward the end of knowledge of nature and the universe, and not only in connection with the service he may render to his neighbor through the use of such knowledge, but also because of what may be learned concerning the Creator in His creation.

## *2. The Importance of Meaning in a Well-lived Life.*

The thoughtful individual who is committed to the exploration and settlement of Mars has such a commitment because it is an extension of, or related to, the source of meaning in his life. Individuals or societies which neglect the importance of meaning breed hopelessness. It was recently noted, "Minds require eco-niches as much as organisms do, and the mind's eco-niche is its worldview, its sense of the whole of things (however much or little that sense is articulated). Short of madness, there is some fit between the two, and we constantly try to improve that fit. Signs of a poor fit are the sense of meaninglessness, alienation, and anxiety that the twentieth century knew so well." (Smith 26) Or as Porpora wrote in his book, *Landscapes of the Soul, the Loss of Moral Meaning in American Life*: "If to know who we are is to know our place in the cosmos, then we cannot lose our place in the cosmos without losing ourselves as well." (152)

For many within our culture, a sense of ultimate meaning has greatly diminished. In part, the sense of meaning has diminished most among those committed to a so-called naturalistic/materialistic worldview. Huston Smith—an expert in comparative religion and worldviews— observes,

> Modernity's Big Picture is materialism or (in its more plausible version) naturalism, which acknowledges that there *are* immaterial things—thoughts and feelings, for example—while insisting that those things are totally dependent on matter. Both versions are stunted when compared with the traditional outlook. It is important to understand that neither materialism nor naturalism is required by anything science has discovered in the way of actual facts. We have slid into this smallest of metaphysical positions for psychological, not logical, reasons. (Smith 20)

Thus, Smith concludes,

> There is within us—even the blithest, most lighthearted among us—a fundamental dis-ease. It acts like an un-

> quenchable fire that renders the vast majority of us incapable
> in this life of ever coming to full peace. This desire lies in
> the marrow of our bones and the deep regions of our souls.
> All great literature, poetry, art, philosophy, psychology, and
> religion tries to name and analyze this longing. We are sel-
> dom in direct touch with it, and indeed the modern world
> seems set on preventing us from *getting* in touch with it by
> covering it with an unending phantasmagoria of entertain-
> ments, obsessions, addictions, and distractions of every sort.
> (Smith 28)

In a culture which increasingly finds itself unable to answer a fundamental question—"How am I to live?," a question which, when
answered, gives meaning to life—we find ourselves without a purpose
which motivates us, marking time trying to numb or amuse ourselves
while awaiting death. Thus Porpora writes:

> The problem in the modern or postmodern world is a per-
> vasive loss of emotionally moving contact with a good that
> is ultimate, a contact that was once provided by the sacred.
> We are still emotionally moved by goods, but they tend to
> be goods that are less than ultimate—family, friends, and
> material possessions. As a consequence, the whole of our
> lives is without any overarching moral purpose. There is a
> lot of talk today about our loss of vision, and that is what
> the loss of overarching moral purpose entails. This hardly
> means that we are all immoral. It does mean that our sense
> of morality has become largely procedural. (71–72)

Porpora's study found that there was a strong correlation between
religious faith and a perception of a meaning to life—an ultimate
meaning to which various purposes in life are related. "Fifty-five
percent of those who say they always think about the meaning of
life also describe themselves as very religious. In contrast, only 22
percent of those who often or sometimes think about the meaning

of life and under 13 percent of those who seldom or never think of the meaning of life also consider themselves very religious." (Porpora 153) Among those people with well-defined religious views, there is a strong correlation with a meaningful life. If one seeks to understand a Christian's motivation for space exploration and colonization, one should look to his understanding of his Creator, and the central virtues of his faith.

## 3. Drawing Inspiration from Three Virtues.

In his first epistle to the Church in Corinth, St. Paul wrote at length concerning the power of love, and its enduring character. The things of this world are passing away, but Paul declared that the virtues endure. As St. Paul declares at the conclusion of chapter 13: "And now abide faith, hope, love, these three; but the greatest of these is love." (1 Cor. 13:13) Let us consider these three virtues in relationship to the Christian's interest in the exploration of Mars.

### 3.1 Faith and the Divine Revelation in Nature.

In the 38th chapter of Job, the Lord asks Job: "Where were you when I laid the foundations of the earth? Tell Me, if you have understanding. Who determined its measurements? Surely you know! Or who stretched the line upon it?" (Job 38:4–5) Again, "Do you know the ordinances of the heavens? Can you set their dominion over the earth?" (Job 38:33) Man's knowledge today continues to be quite limited—from the subatomic mysteries of quantum physics, to the mysteries of the deep reaches of the heavens, there are many theories (and much which is taken for granted) and still little which is *known*. Again, in the words of Isaiah 40: "It is He who sits above the circle of the earth, and its inhabitants are like grasshoppers, who stretches* out the heavens like a curtain, and spreads them out like a tent to dwell in. He brings the princes to nothing; He makes the judges of the earth useless." (Isa. 40:22–23) The understanding that

---

* "Stretches"—present tense; interesting in an expanding universe!

82

man is relatively insignificant ("like grasshoppers") well summarizes man's place before the One who "stretches out the heavens."

The study of nature is based on the non-verifiable epistemic principle that we can meaningfully know anything about nature. Science (as we now know it) arose at the time and place that it did because it rested on the belief that there is a Creator who can be known, and whose handiwork can be studied by all. In the words of the Psalmist, "The heavens declare the glory of God; and the firmament shows His handiwork. Day unto day utters speech, and night unto night reveals knowledge. There is no speech nor language where there voice is not heard. Their line has gone out through all the earth, and their words to the end of the world." (Psa. 19:1–4a) A stunningly aspect of these statements is the accessibility of the universal language of mathematics and information: the very heavens, the psalmist declares, contain information which is universally accessible: "There is no speech nor language where their voice is not heard." Such *natural knowledge* (to speak in theological terms) is available to all and it bears witness to the Creator and the creation.

The Christian advocates space exploration because he want to know more about this creation which the Triune God has made and declared to be "good" (Gen. 1:31). To be a follower of Jesus Christ is to be one who loves truth; Jesus declared, "If you abide in My word, you are My disciples indeed. And you shall know the truth, and the truth shall make you free." ( John 8:31–32) The Christian seeks truth both in *revealed* (that is Scripture) and *natural* knowledge of the Creator and His creation.

### 3.2 Hope and the Future of Man.

Following the virtue of *faith* is that of *hope*. The hope which is central to the Christian faith is the rooted in the incarnate Son of God, Jesus Christ, and the salvation which He established through His suffering, death, and resurrection from the dead. The hope which He gives to believers gives meaning to this life, and the promise of eternal life in and with Him. This death-transcending hope markedly differs from the hopelessness and purposelessness of many within our culture. As Smith observes,

> As for the scientific worldview, there is no way that a happy ending can be worked into it. Death is the grim reaper of individual lives, and whether things as a whole will end in a freeze or a fry, with a bang or a whimper (or keep cranking out more insentient matter in an expanding universe) is anybody's guess. Teilhard de Chardin tried heroically to introduce teleology into the universe with his Omega Point, but his vision has passed neither theological nor scientific muster. Theologians want to know where the 'fall' and crucifixion are in his scenario, while scientists are downright contemptuous. (Smith 37)

Christianity gazes into—and beyond—the end of this fallen creation, trusting in the Creator who infinitely surpasses His creation, even as He cares for His creatures. We thus hear the words of Psalm 102:

> Of old You laid the foundation of the earth, and the heavens are the work of Your hands. They will perish, but You will endure [continue]; yes, they will all grow old like a garment; like a cloak You will change them, and they will be changed. But You are the same, and Your years will have no end. The children of Your servants will continue, and their descendants will be established before You. (Psa. 102:25–28)

Therefore, for the Christian, there is no contradiction between an awareness of the limits of man—and the finite span of his days—and an active, meaningful life in the present. The Christian lives in the awareness of all human civilizations being transient; "For here we have no continuing city, but we seek the one to come," as the writer to the Hebrews (13:14) proclaims. The awareness that this world is passing away is not a grounds for despair, but a spur to using time and abilities to the glory of the Creator. As St. Peter wrote:

Therefore, since all these things will be dissolved, what manner of persons ought you to be in holy conduct and godliness, looking for and hastening the coming of the day of God, because of which the heavens will be dissolved, being on fire, and the elements will melt with fervent heat? Nevertheless we, according to His promise, look for new heavens and a new earth in which righteousness dwells. (2 Peter 3:11–13)

Therefore, establishing colonies on Mars is not—at least for the Christian—motivated by a panicked desire to save humanity by spreading him among the stars. (Space colonization is an astonishing vanity if one believes humanity and the universe lack teleological significance in the face of the eventual heat-death of the universe.) Rather, Christians would establish (or at least live in) such colonies because we seek to share the hope which is ours wherever mankind may be found, and we live out lives which have meaning and purpose as we live in the expectation of the life of the world to come.

### 3.3 Love and the Desire to Serve the Neighbor.

St. Paul lists love as chief among the virtues under consideration: "And now abide faith, hope, love, these three; but the greatest of these is love." (1 Cor. 13:13) Such love is not a fuzzy sentimentality: it is a love which is so directed toward the "other" as to be willing to sacrifice oneself for the "other." As Jesus told His disciples on the night when He was betrayed: "Greater love has no one than this, than to lay down one's life for his friends." (John 15:13)

The hard realities of the life of Mars colonists necessitates that settlers be prepared for self-sacrifice for the sake of one's neighbors—one's friends. The selfishness which views others in terms of what they can do for me, rather than how I might best serve them, is a vice which Mars cannot afford either sociologically or spiritually. Mars will offer ample opportunities for the exercise of such self-sacrificial love.

Enumerating these virtues is not an attempt to impose such values on others, nor is it intended to be an all-inclusive list of the motivations which stir a Christian's interest in the exploration and colonization of Mars. But it should be recognized that wherever humanity spreads, men and women of faith will be present, seeking to serve God as they serve their neighbors.

## 4. Recapitulation and the Cosmos.

In conclusion, a few words ought to be said concerning the doctrine of recapitulation; that is, the biblical teaching which stresses Christ as the 'New Adam,' the Head of redeemed humanity. Thus, for example, St. Paul sets forth the comparison between Adam and Jesus in Romans 5: "For as by one man's disobedience many were made sinners, so also by one Man's obedience many will be made righteous." (v. 19) And thus St. Paul presents the creation itself awaiting this new creation in Christ being brought into its fullness: "For the earnest expectation of the creation eagerly waits for the revealing of the sons of God. For the creation was subjected to futility, not willingly, but because of Him who subjected it in hope;" (Rom. 8:19–20).

The doctrine of recapitulation is strongly emphasized in the profoundly influential writings of the second century Church father St. Irenaeus. But the teaching has much to inform Christians in their whole relationship with the creation. Man's responsibility as a steward—not master—over creation is connected to the understanding that Christ is Lord over heaven and earth and men must give an account of their stewardship. On Mars we may seek to serve God through serving our neighbor, using the good creation of God to His glory, possibly bringing life to a dead world, bringing hope to the suffering, and meaning which glorifies the Creator.

# VIII.

## Observations Regarding the Intellectual, Ethical and Spiritual Formation of Future Martians.*

When Robert Zubrin's *The Case for Mars* was published back in 1996, the present author, along with many other readers, was struck by Dr. Zubrin's ability to see the 'big picture' ; here was a vision where the goal of colonizing Mars is not just about technological accomplishment; it is a concern for the preservation of human civilization. As Zubrin observed in his epilogue, "The Significance of the Martian Frontier":

> Currently we see around us an ever more apparent loss of vigor of our society: increasing fixity of the power structure and bureaucratization of all levels of life; impotence of political institutions to carry off great projects; the proliferation of regulations affecting all aspects of public, private, and commercial life; the spread of irrationalism; the banalization of popular culture; the loss of willingness by individuals to take risks, to fend for themselves or think for themselves; economic stagnation and decline; the deceleration of the rate of technological innovation. ... Everywhere you look, the writing is on the wall.
>
> ...
>
> The creation of a new frontier thus presents itself as America's and humanity's greatest social need. Nothing

---

*This essay was originally presented at the 2000 convention of the Mars Society in Toronto, Canada.*

is more important: Apply what pallatives you will, without a frontier to grow in, not only American society, but the entire global civilization based upon values of humanism, science, and progress will ultimately die. (Zubrin 297)

For Zubrin and many others, opening the Martian frontier is seen as necessary for the preservation of Western civilization. Certainly the societal problems Zubrin identifies are among the signs of decadence and decay which are arising. However, when one searches for the causes of cultural decline, they are far more profound than the simple absence of a frontier. As important as a 'new frontier' might be, it is an inadequate solution to the problems which confront our civilization.

The great revival of knowledge and faith fueled by humanists and reformers began in the early fourteenth century—well before the re-discovery of the 'New World'—and commenced through much of its most fundamental development in seeming ignorance of the possibilities posed by an open frontier. Remarkable figures of the age, such as Petrarch, Erasmus of Rotterdam, and Philipp Melanchthon of Wittenberg would have aided in the reshaping and revitalization of their culture regardless of the existence of a new frontier. Theology, philosophy, literature and the other arts, and science were all revitalized in the old world and then exported to the new world. It is true that the Americas provided a virtual tabula rasa for exploring the renewed vigor of Western thought, but the frontier did not initiate the renewal of civilization; rather, it provided an outlet for its expression.

Thus, if the purpose for going to Mars is for permanent colonization, the spiritual, ethical and intellectual formation of future 'Martians' should be a matter of intense interest. A Martian frontier, like the frontiers which preceded it, will give birth to 'new' ideas to the extent that such ideas are only partially fulfilled already in the 'old world.' In the words of Horace: "*Caelum non animum mutant qui trans mare current*"—"They change the sky, not the soul, who cross the ocean." (cited in Cahill 193) Utopian fantasies are useful insofar as they help a culture explore its core values; they are troublesome, even dangerous, if they give rise to the idea that men will suddenly be spiritually transformed simply by changing location. Different environments may be seen as opportunities for expressing interests or using knowledge which previously lay dormant, but it does not fundamentally alter who you are. The sky changes, but not the soul.

This realization has profound implications for our endeavors. Our role here is not to in any way diminish the importance of the many fine papers which have been, or will be, presented on crucial scientific and technical aspects of traveling to Mars and opening the new frontier. Instead, it is the aim of the author to sound a note of caution, lest we neglect the equally important matter of preparing men and women who are truly ready for such a task. The focus of this paper is not on the scientific and technical training of potential first generation settlers, but concerns more fundamental aspects of their educational and spiritual formation.

It is understandable why many may view settling Mars as an opportunity to "start over." Certainly the realistic view Earth-side is bleak if one considers the future of civilization. Malcolm Muggeridge wrote in his essay, "The Great Liberal Death Wish,"

> As the astronauts soar into the vast eternities of space, on earth the garbage piles higher; as the groves of academe extends their domain, their alumni's arms reach lower; as the phallic cult spreads, so does impotence. In great wealth, great poverty; in health, sickness, in numbers, deception. Gorging, left hungry; sedated, left restless; telling all, hiding all; in flesh united, forever separated. So we press on through the valley of abundance that leads to the wasteland of satiety, passing through the gardens of fantasy; seeking happiness ever more ardently, and finding despair ever more surely. (cited in Kirk 5)

Unlike the committed humanists and reformers whose efforts were the means by which our Western civilization rose to global prominence this last half millennium, there are many of their heirs who elect to squander their inheritance. Such atomized individuals have virtually ceased to exist as true members of society because they view their relationships in terms of personal advantage, rather than mutual obligation and responsibility. They have ceased to understand that they exist as individuals through their relationships with others.

What kind of Mars do we want? The first generation of permanent settlers will have a defining role in shaping Martian culture for generations to come. When you lay a foundation, you determine the future of the whole house. An inadequate foundation dooms the building before the structure is even finished. Therefore, there are aspects of character beyond choice of vocation which are crucial in settlers. More than simply technical competency is required. It is not enough that such settlers are the "best and the brightest"—they must be more than simply intelligent; they must be virtuous. St. Paul wrote to the Romans, "And though I have the gift of prophecy, and understand all mysteries and all knowledge, and though I have all faith, so that I could remove mountains, but have not love, I am nothing." (Romans 13:2) If Mars is settled by individuals who care for nothing more than themselves, if they are loveless and self-obsessed, then we would do just as well to stay home, for the endeavor will come to nothing. Indeed, in such a situation one cannot help but think of the failure of the Jamestown colony in seventeenth century Virginia, where self-obsessed fortune seekers did not gain fortune and even forfeited their lives.

Jesus declared, "Greater love has no one than this, than to lay down one's life for his friends." (St. John 15:13) Self-serving escapists lack the most crucial element of virtue: the willingness to sacrifice—whether in terms of comfort, wealth, or even life—for the benefit of others. Settlers must be firmly established in such a loving ethical standard because many of the challenges they will face will test their integrity. No one should suffer any utopian illusions as we contemplate the exploration and colonization of Mars—men and women will never have attempted to settle such a desperately hostile environment. In a crisis, they may need to being willing to die for one another; they must see beyond their narrow interests to embrace the needs of others. Such a environment does not allow for the morally bankrupt to continually put themselves and their personal gain before their society.

St. Augustine wrote: "In order to discover the character of any people we have only to observe what they love." (cited in Codevilla 28) Two great loves, therefore, must characterize those who will form the new civilization: love of neighbor and love of the truth. Both of these loves, however, are notably counter-culture today, for atomistic individualism and cultural relativism are formative influences of our times.

The self-centered individual is like the lawyer in St. Luke's Gospel, who, "wanting to justify himself, said to Jesus, 'And who is my neighbor?'" (10:29) And yet he had already admitted that the two great commandments of the law are "You shall love the LORD your God with all your heart, with all your soul, with all your strength, and with all your mind, and your neighbor as yourself." (St. Luke 10:27) The self-centered individual is unwilling to see each neighbor's need as if it were his own need, and thus he seek to limit the circle of 'neighbors' as tightly as possible.

The blight of cultural relativism is equally damaging to civilization, for it assaults truth itself. As Allan Bloom observed, "Yet everyone likes cultural relativism but wants to exempt what concerns him. The physicist wants to save his atoms; the historian, his events; the moralist, his values. But they are all equally relative. If there is an escape for one truth from the flux, then there is in principle no reason why many truths are not beyond it; and then the flux, becoming, change, history or what have you is not what is fundamental, but rather, being, the immutable principle of science and philosophy." (Bloom, 203) It is the heart of the cultural relativist which allows Pontius Pilate to look into the face of Him who is the Way, the Truth and the Life and ask, "What is truth?" (St. John 18:38) A cultural relativism which aims at attacking the universality of moral or ethical truth claims ends up eliminating *all* truth claims; this is the "spread of irrationalism" which Zubrin laments. (297)

The opening of the new frontier requires men and women who reflect a renewed 'humanist' spirit. (We mean by this the genuine Renaissance humanism which opened and expanded minds, not its banal secular 'knock-off' which closes minds and welds them shut.) Ours is an age of narrow specialization when men and women are expected to know a great deal about very little. The colonization of Mars requires educated men and women capable of understanding our accumulated history and experience outside of narrow parochial perspectives. Barzun describes the humanists as those whose "continuing enthusiasm for the ancients was reinforced by the feeling that the inherited culture was dissolving and here was a storehouse of ideas and attitudes with which to rebuild. It was like going up to the attic and polishing up semi-discarded treasures." (Barzun 46) We, too, must learn to treasure the experience of history; the ancient treasures must be brought out and polished again.

In short, prospective settlers need to be 'educated' in the classic, liberal sense of the term. Such a liberal education is not in opposition to scientific training; rather, the first is foundational to the latter. Unless one's specialization is preceded by a grounding in the wealth of human knowledge, the spirit is blunted and dull. As Dean Briggs of Harvard College remarked a hundred years ago: "The new degree of Bachelor of Science does not guarantee that the holder knows any science. It does guarantee that he does not know any Latin." (cited in Barzun 45) Being truly educated is of service in any vocation and is of lasting benefit to the culture whether or not one goes to Mars. In the words of James Russell Lowell, the study of the classics "is fitly called a liberal education, because it emancipates the mind from every narrow provincialism, whether of egoism or tradition, and is the apprenticeship that every one must serve before becoming a free brother of the guild which passes the torch of life from age to age." (cited in Kirk 42) If a fruit of the Apollo program was numerous scientists and engineers, perhaps humanity will realize even greater gains from a generation which can revitalize Western culture. Again, in the words of Russell Kirk:

> The more people who are humanely educated, the better. But the more people we have who are half-educated or quarter-educated, the worse for them and for the republic. Really educated people, rather than forming presumptuous elites, will permeate society, leavening the lump through their professions, their teaching, their preaching, their participation in commerce and industry, their public offices at every level of the commonwealth. And being educated, they will know that they do not know everything; and that there exists objects in life besides power and money and sensual gratification; they will take long views; they will look forward to posterity and backward toward their ancestors. (Kirk 46–47)

It seems virtually indisputable that our age is one of decadence and decay in Western culture; for some, this is a cause for rejoicing, but for many it brings confusion and sadness—a sense of having lost something which is almost indescribable. In efforts to preserve that which is worth saving in our culture, we must remember that, ultimately, culture arises from the *cultus*—a society's foundational beliefs. The shared faith (or lack thereof) will profoundly shape a society. Many of the ills undermining present Western civilization are intimately linked to the decline of the sacred. As Codevilla observes,

> The most contentious and consequential issues touch religion. Modern government's relationship with religion has been one of rivalry. Although there is not now and never has been a better predictor of prosperity, family, and civility than the practice of Judaism and Christianity, modern Western governments have used their power over education to teach secularism at first, followed by various anti-religious dogmas and, most recently, lifestyles repugnant to religious

morality. The fundamental Judeo-Christian teaching is that mankind lives under a single, objective set of laws equally binding on all. As governments drain Western societies of religious preferences, they introduce new beliefs based on relativism, that is, on the basis of power. Hence, nowadays nihilism does battle for Western souls with a thin, ill-fitting combination of self-worship and earth worship. (Codevilla 14)

Any future Martian culture must witness a restoration of the role of the sacred within the lives of its people, or it will share the spiritual ills of our present culture. A renewed culture, rooted in love of neighbor and zeal for the truth—especially zeal for the One who is Truth—will be truly ready for the new frontier. Let us grow in our souls, first, and then set out upon the sea.

# IX.
# Faith, Culture and Community in a Future Martian Civilization.

W
hat kind of Mars do we want for our descendants? Science fiction presents us with a Mars of many colors: the Red-Green-Blue Mars of Kim Stanley Robinson, the *White Mars* of Aldiss and Penrose, even the *Rainbow Mars* of Larry Niven. In *Red Dust* Paul Mcauley presents us with a Chinese Mars; in *Crescent in the Sky* and *A Gathering of Stars*, Donald Moffitt gives us a Moslem Mars; Sigmund Brouwer, a Christian Mars. Beyond the rainbow of our multicolored models for Mars are a wide variety of answers to a fundamental question: What kind of Mars do we want for our descendants? Each author presents a vision for the future. This is a question, ultimately, of culture. The realm of science fiction functions as a 'simulator': it provides a forum for modeling possible futures, providing portraits of varying degrees of plausibility of cultures which could come to be. The degree to which a particular story inspires or interests us is not just a factor of scientific plausibility—some novels prove that good science alone does not make a good story. Most successful stories tell us of people living in community.

In the fourth century B.C., Aristotle observe in his Politics: "Every state is a community of some kind, and every community is established with a view to some good; for mankind always acts in order to obtain that which they think good." (Book I.1) The "good" toward which each community is oriented rests at the center of its

---

*This essay was originally presented at the 2005 convention of the Mars Society in Boulder, Colorado.*

culture. A culture is the sum of the attitudes, customs and beliefs that distinguish a particular group of people. Culture and *cultus* cannot be separated: that which a community worships or venerates forms the whole pattern of beliefs, customs and values of that community. The Latin word *colo* (which can mean "to inhabit," "to cultivate," "to till" and "to worship") is the root for both "culture" and "colonize". In our language, inhabiting a place and tilling its land is conceptually connected with worship.

It is not a coincidence that a decline in the Christian *cultus* of the West has been accompanied by a change in the way in which mankind relates to the earth and within human community. As elements of Western society have consciously (and sometimes unconsciously) become disconnected from their Christian roots, man has adopted a utilitarian materialistic outlook in his relationship with the earth and the rest of nature which, rather than viewing himself within a relationship as steward to the creation, views the rest of the natural world and one's fellow human beings, as merely resources to be exploited. As Thomas Hughes observes in his recent book, *Human-Built World*: "Seeing a human-built world around them, engineers, scientists, and managers believed that they had the creative technological power to make a world according to their own blueprints. They considered the natural world as expendable and exploitable or simply as scenery." (9) Hughes might be chided for his 'insensitivity' toward the feelings of such engineers, scientists, and managers, but his assessment is precisely correct. Supposing themselves freed from objective ethical constraint, in that they view morality as simply another human 'construct,' utilitarian materialists increasingly demand that the few remaining ethical principles restraining their conduct be removed. Certainly not all of those individuals within these fields subscribe to the new worldview which is endeavoring to dominate the culture; in fact, possibly even a majority of those working within such specializations still adhere to the received culture to one extent or another. The problem is that ethics, philosophy, and theology (disciplines concerned with qualitative, rather than quantitative properties) are not their field of specialization, and the proponents of the material-ist dogma have attempted to define the qualitative disciplines out of existence. Thus Hughes observes:

> Technologically empowered, we have reason to doubt our values and competence as creators of the human built and as stewards of the remaining natural world. Slums in inner cities, ugly strip malls in the suburbs, hastily and cheaply built housing, polluted air and water, the loss of ecologically nurturing regions, and the likely threat of global warming give evidence of our failure to take responsibility for creat-ing and maintaining aesthetically pleasing and ecologically sustainable environments. (Hughes 153)

Without the received *cultus*, man is uprooted; he no longer is tied to a place and a community. Alienated from other human beings and from the rest of the natural world, his activities become destructive of community and nature. Exeter University Professor Tim Gorringe observes in his book, *A Theology of the Built Environment*:

> To be human is to be placed: to be born in this house, hospital, stable (according to Luke), or even, as in the floods in Mozambique in 2000, in a tree. It is to live in this council house, semidetached, tower block, farmhouse, mansion. It is to go to school through these streets or lanes, to play in this alley, park, garden; to shop in this market, that mall; to work in this factory, mine, office, farm. These facts are banal, but they form the fabric of our everyday lives, structuring our memories, determining our attitudes. How, as Christians, should we think of them? ... Paul constantly urges his congregations to 'edify' one another. The word 'edify' comes from the Latin *aedificare*, to build. The metaphorical use of the word points to a profound truth about the built envi-ronment. Form follows function; buildings serve a purpose. For good or ill buildings, from the humblest garden shed to the grandest cathedral, make moral statements. (Gorringe 1)

Being "placed"—linked to the land and 'putting down roots' according to the common expression—is fundamentally a part of being human. Therefore, the connections implicit in colo, the root of "colonize," "culture," and *cultus* —that is, "to inhabit," "to till," and "to worship"—has clear implications to efforts to settle human communities on Mars. Such Martian communities may have a shared culture, or they may have many discrete cultures. What is clear is that where there is a community, there will be a common "good," a shared set of values, rooted in veneration in a common *cultus*. As Gorringe observes, the culture will be expressed within the community even in the physical expressions of community, through architecture and the expression of the balance between 'work' and 'home':

> The built environment, which 'provides us with all the most direct, frequent and unavoidable images and experiences of everyday life', is never just happenstance. It reflects conscious decisions which in turn reflect ideologies and class positions. 'Grasped as an image', says Heinrich Rombach, 'the basic character of a farmhouse says a great deal more about the "spirit" of the country, and a style of building reveals more of the basic philosophy of a period, than the carefully smoothed-out texts of the school philosophy of the time.' Not just farmhouses, we have to add, but council estates, tower blocks and out of town shopping centres; and not just philosophy, but theology. Theology, as one form of ideology, plays its part in the shaping of space, and not just in overtly religious buildings, nor just in pre-secular societies. (Gorringe 8)

The truth of Gorringe's observation may not be readily recognized by those whose experiences have only been limited to a single culture. Nevertheless, experience within the broader human context—especially if one has the opportunity to visit a completely different culture—readily substantiate its accuracy.

The significance of Gorringe's observation regarding the relationship between the 'human built' world and human culture is clearly evident in discussions concerning the possibilities for a future community (or communities) on Mars. And yet to date nearly all discussion of potential human settlements has focused on engineering issues, rather than the even more central issues of community. If to be human is to be placed, what will it mean to a human community to be placed on Mars? What will be the common "good" of such a community, or communities? The question is inescapable—and it is a question which will be answered "on the ground," even if there was not much thought given to the matter in advance. Why, then, not simply leave the question to the Martians? Because such a course of action would truly fulfill the cliché about 'putting the cart before the horse': the motivating "good" pulls the society; it logically precedes the society which derives from it, providing an organizing principle which finds expression in a multitude of ways throughout the whole of the culture. Again, Gorringe observes:

> A community is a group of people who have something in common, but what? Throughout history this has included the idea of territory, and territory, I will argue, remains important, even in an age of migration and of identity politics. However, territorially based notions of community are challenged by class, caste, gender, and ethnic divisions because community is also rooted in religion, culture, language, ethnicity, work or ideas. That idea that it is not territory which has primacy but shared interests and values is not an invention of the twentieth century but was alive in first century Galatia, and indeed in Egypt under the Pharaohs. Hardly surprising then that 'alternative' communities go back at least to the Essenes in the first century, and run through Benedict, Bernard of Clairvaux, Francis and Dominic, the Anabaptists, and Robert Owen, to the communes and base communities of the late twentieth century. All of these

represent attempts to find the way back to the Garden of harmonious community. (Gorringe 163)

Some might respond that the common "good" is settling Mars, but I believe that in almost all cases such a response actually misses the mark. It prompts the question, "But why settle Mars?" The answer to this more fundamental question, I believe, will get nearer to the motivating "good" of settling Mars.

Interestingly, much of the science fiction concerned with the possibility of human settlement on Mars deals with the cultural questions far more often that one finds to be the case in the more scholarly literature. I suspect that the relative dearth of attention paid to cultural questions regarding the possibility is the result of at least two major factors. First, most of the individuals involved in thinking through the possibilities for establishing human communities on Mars come from either an engineering or scientific background; they will seek to apply their areas of competence to the problem at hand, as is to be expected.

The second major factor is a hesitancy to speak of cultural matters at all, in part because of the attempt to collapse the qualitative into the quantitative—an effort which consistently ends in failure. As Stanley Jaki observes:

> Faced with that inability [to explain all qualitative matters by quantitative means], the scientist can take two attitudes. One rests upon the mistaken conviction that the scientific method is everything and whatever cannot be expressed in quantitative terms, is purely subjective, that is, illusory. Such was, for instance, the attitude of Einstein, who said that consciousness and free will are no objective realities, because they cannot be handled by physics. He might as well have called them sheer illusions. Clearly, it is better to take another attitude and acknowledge that there are some basic limits to a limitless science. Those limits appear as soon as a question arises that cannot be put in a quantitative form

and therefore cannot be given a quantitative answer to be tested in a laboratory. (11)

Again:"Human knowledge, whether we consider it to have come from the hand of God or not, concerns two separate realms, quantities and non-quantities, and these two realms are irreducible to one another. It is not profitable for man to chafe under that restriction." ( Jaki 19)

To a certain degree, it might well be that we are now reaching a certain understanding of what life on Mars, quantitatively-speaking, will be like. Although the engineering details of such habitations are, of course, far from complete, and our knowledge of the geology and hydrology (for starters) is also far from complete, one can draw up a list of the known physical needs of a settlement, and begin to offer technical solutions to such material needs on the basis of *in situ* resources. It is the qualitative which is far from determined, and it is the qualitative which is, by far, the more important. Quantitatively, the settlement will adapt to the situation as they find it; qualitatively, the settlers will bring with them that which will mold them as they encounter a new world, and even though their worldview will adapt to the circumstances, the initial condition will be of tremendous significance. As Jeffrey Kargel recently observed:

> Whether the character of human beings and the behavioral pattern of the future Mars society will materially diverge constructively from Earth's history may depend on choices made in selecting the first permanent Mars crews. This will be essentially Earthlings' last big choice that will direct the New Martians' future. After that, it will be pretty much up to the New Martians to make of it what they will. It will be interesting to see how this crew selection process unfolds and how it may affect the political and sociological foundation of the new Mars society. Will we seek a broad representation of humanity to venture forth to Mars? Will it be a diverse crew or an ethnically, religiously, politically, and culturally monolithic crew? Will democracy prevail or

will a centralized authority attempt to impose a common good? What tolerances will there be for individual behavior and allowances for personal achievement and reward versus communal work for the public good? Will the concept of nationalistic pride become extinct? How will suicidal, murderous, and terroristic tendencies be dealt with, or will these behaviors be successfully selected against right from the start? Will class distinctions be eliminated right from the start, or will it be incorporated into the economic structure of a society based on free enterprise and meritorious reward? Will bigotry be defeated on Mars and left behind on Earth? (434)

Although Kargel does not provide much by way of what he would prefer be the answer to these (and many other) questions, he is to be commended for asking the questions. In coming to an answer to these questions, there is a virtue in the necessity that we will almost certain 'start small' when it comes to settlements. The first permanent colony may only be one or two dozen individuals; certainly it would not amount to more than a hundred souls. I strongly suspect that initially the wrong criteria will be utilized: there will be an overemphasis on the *curriculum vitae*, and the result could well be a colony of egocentric political manipulators, such as the first hundred one encounters in Kim Stanley Robinson's *Red Mars*. Hopefully Earth will not squander the opportunity to begin again; perhaps we will wisely allow for a variety of communities which will explore their own conceptions of the "good." I suspect that such diversity, not unlike genetic diversity, is good for the species. At the very least, pursuing the course of a variety of communities will avoid recklessly discarding options before the Martians have the opportunity to find their own way from the options at hand. Again, in Kargel's words:

> Most people in the developed world probably recognize Mars as part of human destiny; this is particularly true in the USA, where a pioneering spirit was a necessity of survival

for several centuries. Exploration of Mars is widely regarded as a noble and pure expression of the best in humanity. We look forward to the detailed exploration for Martian life; to the elucidation of the causes and processes of global climate change and derivation of lessons for Earth; to other types of science that could be done on Mars, for instance astronomy done with ultra-long-baseline radio telescope arrays. We look forward to the technical challenge and the communion of human spirit that will develop as we go on to Mars and establish a sustainable, growable presence there. (Kargel 476)

Thus Kargel weds the technical challenge (the quantitative) and the "communion of human spirit" (the qualitative) which make a civilization, a culture. The pursuit of the technical challenge will be shaped by, and will in turn influence, that communion of human spirit. The Martians will, I suspect, pursue an agenda all their own, and yet it will be one which, from the retrospective of another thousand years (if the Lord so wills) seem an obvious extrapolation from the foundation which was laid.

As with all new beginnings, Mars offers a very rare opportunity to learn from our failures and, even though never starting from a blank slate (mankind is never a *tabula rasa*), it is an opportunity to avoid repeating some of our past and present mistakes. Commenting on the plights of our modern culture, Gorringe observed:

> Though the quest for community is ancient, it assumed a new intensity in the wake of industrialisation, which has involved, amongst other things, the dissolution of the small rooted communities which human beings had lived in for millennia, and their replacement by the anonymous city. These changes were theorised by Frederick Tönnies in 1877 in terms of the distinction between *Gemeinschaft*—roughly, the community of the village, based on mutual aid and trust, and centred on family, neighbourhood and friendship—and *Gesellschaft*, the association of the anonymous city, based on

individual self interest and contract. He believed that true
community was only possible in the former. (Gorringe 164)

What kind of Mars do we want for our descendants? The goal should
be *Gemeinschaft*, true community, in pursuit of the common good. The
goal should be that those who inhabit a new world, tilling the soil together, would gather in veneration of that which is good. We are told
by St. James, "Do not be deceived, my beloved brethren. Every good
gift and every perfect gift is from above, and comes down from the
Father of lights, with whom there is no variation or shadow of turn-
ing. Of His own will He brought us forth by the word of truth, that
we might be a kind of firstfruits of His creatures." (James 1:16–18)
Having turned away from the One who is Good and the Truth, there
are many who are left in doubt today as to what is good and true. As
every community is centered on pursuit of the common "good" there
can be no greater center than veneration of the source of all good.

Our record, in this fallen world, is not a good one. In his
insight book, *Collapse*, Jared Diamond, having traced the factors
that have led to the collapse of many civilizations and which now
threaten our own, observes, "Our world society is presently on a non-
sustainable course..." (498) The recurrent point throughout his book
is that the problem is not one of engineering, but of the heart and
mind—a matter of spirit. Our societal ills rest in the qualitative, not
quantitative; our thoughts for a Martian future must center there, as
well. Then, with eager steps and true confidence, we may mark out
our steps in the red sands.

# X.
# Examining the Prospects for a Red Planet Renaissance.

## Introduction—What Kind of New World? Two Examples from the Past.

To the millions afflicted by the intellectually blinding plight of historical myopia, the course of the opening and development of the Western Hemisphere appears inevitable—the only conceivable outcome to the discovery of the New World. The reality, of course, was far different; different cultures struggled to imprint their image on the newly discovered lands. The culture of the New World grew out of the struggles of the Old World. The borders of the New World often grew out of victories on the battlefields of Europe. Conflicts between the Dutch, Swedish, English, French and Spanish governments before, during, and after the Thirty Years War established and moved boundaries.

A central example of the Old World influence can be found in events leading up to Columbus' famous discovery. In *The Case for Mars*, Dr. Robert Zubrin observed, "People everywhere today remember Ferdinand and Isabella only because they are associated with the voyage of Christopher Columbus." (294) Certainly this is a correct assessment of popular historical knowledge, but this does not mean that funding Columbus was their most important act. Spanish history in the late 15<sup>th</sup> century revolves around the conclusion of the *Reconquista*, the final expulsion of Islamic forces from the

---

*This essay was originally presented at the 2003 convention of the Mars Society in Eugene, Oregon.*

Iberian peninsula after 780 years of occupation. As one recent author observed:

> Ferdinand and Isabella brought modern Spain into existence. The unification of Spain was the main achievement of the royal couple. It was accomplished through the defeat of the Jihad. The human toil of this victory was horrendous, but in view of the divisive nature of Islam, to battle it was the only course acceptable to the Spanish nation. There would have been, perhaps, one other route to unity: the Islamization of Spain. That was the Jihad way. But Spain would no longer have been itself; another country would have emerged, perhaps flourished, as a European extension of North Africa. (Fregosi 270)

On January 2, 1492, the final Islamic enclave surrendered to Ferdinand's forces. At the capitulation ceremony, "One of the people present ... was a Genoese navigator by the name of Christopher Columbus, who was seeking aid from Queen Isabella for a voyage he wished to make west, across the Atlantic to the Indies." (Fregosi 273) Only during the reign of Ferdinand and Isabella did Spain achieve the political and cultural unity that made it possible to fund explorations and, ultimately, establish an empire in the New World which offered one of several competing visions of the future of the Western hemisphere.

However, the New World empire established by Spain reflected the culture of Spain: It reflected the history of centuries of war; it reflected the ambitions of Charles V's dream of universal empire; it reflected the ideology of the Inquisition. The Spanish empire exemplified a culture in which the Renaissance and Reformation were treated with contempt and persecution. New World colonies were viewed primarily in terms of exploitable resources (especially gold and silver) which were to be extracted for the benefit of Spain, rather than utilized for developing a civilization in the Western Hemisphere. Portuguese colonies reflected a similar world view.

The North American colonies established by the French, Dutch, Swedish and English were of a different character. The North American colonies never proved to be of great economic value to the mother countries (in fact, several early English colonies which were established for economic purposes were colossal failures). As one historian observed, "Barbados or Jamaica alone—or, for that matter, some other islands in the West Indies—seemed worth more to London than all the thirteen mainland colonies put together." (Kirk 303) The success of these colonies was not rooted in economics; it was based in the beliefs and values of the settlers, and those who supported them. The civilization which they built was therefore of a markedly different character from colonies rooted in different values and concerns, despite the tremendous material advantages which other colonies possessed.

Obviously, exploration and colonization occurs within a historical context and the exploration of the Moon and Mars thus far has proven that things will be no different in outer space. As we examine the possibility of a "New Frontier" on Mars, an important question for us is: "What kind of Mars do we want?" Or to phrase the question more precisely: "What kind of Martians do we want?" Will settling Mars simply export current problems inherent in our declining culture, or will it offer prospects for a new civilization? The European settlements in the Western Hemisphere gave expression to some of that which was best (and worst) in the 16th and 17th century cultures. Prospects for a "Red Planet Renaissance" should be a matter of central concern for those who desire to open a "New Frontier" because the implications of such a renewal of society benefit not only hypothetical future cultures, but are of immediate value to our own culture.

### The Renaissance and the Opening of a Hemisphere

As the author observed in a paper presented to the third convention of the Mars Society,* the revival of knowledge and faith fueled by humanists and reformers began as early as the 14th century;

*See the preceding essay.

rather than the existence of a frontier bringing on the development of humanism, the men who were shaped by the Renaissance and Reformation applied what they had learned in their age of the renewal of knowledge to the 'clean slate' of a New World. Theology, philosophy, literature and the other arts and sciences were all revitalized in the Old World and then exported to the New World. The English colonies, in particular, provided an opportunity for advocates of various reforms to realize some of their beliefs in practice. As Zubrin observed in *The Case for Mars*:

> The essence of humanist society is that it values human beings—human life and human rights are held precious beyond price. Such notions have been for several thousand years the core philosophical values of Western civilization, dating back to the Greeks and the Judeo-Christian ideas of the divine nature of the human spirit. Yet they could never be implemented as a practical basis for the organization of society until the great explorers of the age of discovery threw open a New World in which the dormant seed of humanism contained within medieval Christendom could grow and blossom forth. (298)

The humanist movement was not a philosophy, but an approach to learning and to sources of knowledge.* A common cry stirred both the humanists and religious reformers: "*Ad fontes!*" For the reformers and humanists, this meant, in part, a return to the study of the Sacred Scriptures and church fathers in their original languages and context. In addition, the humanists were convinced that deliverance from their "dark age" was to be accomplished, in part, by studying the ancient Greek and Roman writers, seeking out their wisdom—not simply by rote repetition, but also by learning from their experiences and by studying their culture and context. The

---

* Thus there is a fundamental difference between authentic humanism, and the ideology of 'secular humanism,' which has attempted to arrogate the humanist legacy to itself.

humanists' faith in classical Antiquity was somewhat naive, but it helped shape a transformation of the Western mind.

> In differing ways and degrees, but always with the dream of creating a better future by capturing the essential qualities of Antiquity, humanists such as Petrarch, Valla, Machiavelli, and Erasmus pioneered in defining the role of the intellectual as conscience, gadfly, critic. ... By the time of Montaigne, and certainly by the seventeenth century, sublime trust in the curative powers of Antiquity had waned; but the critical spirit, the hope of improvement (no longer just by 'rediscovering' Antiquity), and even the habit of using the press to form and appeal to public opinion survived, all of them legacies from the culture of Renaissance humanism. (Nauert 215)

Or as Jacques Barzun recently observed, the humanists were those whose "continuing enthusiasm for the ancients was reinforced by the feeling that the inherited culture was dissolving and here was a storehouse of ideas and attitudes with which to rebuild. It was like going up to the attic and polishing up semi-discarded treasures" (46).

The revival of that knowledge which had been forgotten— and the publication and study of that knowledge—revitalized the culture. The efforts of humanists and reformers to disseminate such knowledge reshaped their world.

Huntington states that "The central elements of any culture or civilization are language and religion." (59) Culture is rooted both etymologically and ontologically in *cultus*. The cultures rebuilt through Renaissance and Reformation were exported to the North American colonies. Massachusetts, Rhode Island, Pennsylvania, and Maryland directly owe their existence to the exportation of such a culture by men and women prepared to leave behind the Old World to live out their ideals on the frontier, and their sacrifices and ideals were emulated by countless millions throughout the generations which followed.

## *The Decline of the West*

History bears witness that the culture of colonizing societies is carried into its colonies. What culture, then, would be carried to Mars? Or to pose the question more directly: Is *our* culture presently prepared to export a healthy culture to a New World? The present status of Western civilization is quite bleak. Carroll Quigley observed that civilizations decline when they stop the "application of surplus to new ways of doing things. In modern terms we say that the rate of investment decreases." (cited in Huntington 303) This comment led Huntington to observe: "This happens because the social groups controlling the surplus have a vested interest in using it for 'nonproductive but ego-satisfying purposes … which distribute the surpluses to consumption but do not provide more effective methods of production.' People live off their capital and the civilization moves from the stage of the universal state to the stage of decay." (303) Certainly such ego-satisfying consumption is rampant within our society, but Huntington observes that there are even graver dangers confronting our culture:

> Far more significant than economics and demography are problems of moral decline, cultural suicide, and political disunity in the West. Oft-pointed-to manifestations of moral decline include:
>
> 1. increases in antisocial behavior, such as crime, drug use, and violence generally;
>
> 2. family decay, including increased rates of divorce, illegitimacy, teen-age pregnancy, and single-parent families;
>
> 3. at least in the United States, a decline in "social capital," that is, membership in voluntary associations and the interpersonal trust associated with such membership;
>
> 4. general weakening of the "work ethic" and rise of a cult of personal indulgence;
>
> 5. decreasing commitment to learning and intellectual activity, manifested in the United States in lower levels of scholastic achievement. (Huntington 304)

In 1976, theologian and philosopher Francis Schaeffer summarized these developments as follows:

> As the more Christian-dominated consensus weakened, the majority of people adopted two impoverished values: personal peace and affluence.
>
> Personal peace means just to be let alone, not to be troubled by the troubles of other people, whether across the world or across the city—to live one's life with minimal possibilities of being personally disturbed. Personal peace means wanting to have my personal life pattern undisturbed in my lifetime, regardless of what the results will be in the lifetimes of my children and grandchildren. Affluence means an overwhelming and ever-increasing prosperity—a life made up of things, and more things—a success judged by an ever-higher level of material abundance. (205)

People who are fixated on themselves—focused on personal peace and affluence, in the words of Schaeffer—do not make the sacrifices necessary to launch a new civilization. Rather, they are more inclined to discard their own civilization (if necessary) to preserve their safety and affluence. Successfully settling a New World requires belief in ideals which transcend the material desires of the individual; a decadent, consumption-oriented society will not long endure as a civilization, let alone give birth to a healthy, new civilization (except through its absence). In the words of Robert Kaplan in his recent book, *The Coming Anarchy*: "Material possessions not only focus people toward private and away from communal life but also encourages docility. The more possessions one has, the more compromises one will make to protect them." (89) Ironically, as a society focuses increasingly on material possessions, it finds that its ability to provide technological advances decreases. Thus Zubrin observes, "Without the opening of a new frontier on Mars, continued Western civilization also faces the risk of technological stagnation." (Zubrin 300)

Lacking frontiers, and the frontier spirit which sought open spaces and opportunities to manifest the ideals of the Western soul, our civilization may soon lose even the ability to continue its relative technological progress. New nations which advance the ideals of a civilization are the engine which give expression to the vitality of that civilization. "Technological progress is cyclical, as is most of history. The West has been privileged to live through two major cycles—the Middle Ages and the Renaissance—within a civilization that has lasted now for a thousand years. The majority of civilizations, like those of Byzantium or Islam, seem to have lived through only one cycle. ... But today the West has no new young nation in reserve, and the momentum cannot be maintained." (Gimpel 240)

## The Prospect of a Red Planet Renaissance

The opening of the Western hemisphere followed the rebirth of civilization in the Old World. Such a rebirth is necessary for the survival of the West and of all that is best in our civilization—it may also bear fruit in the expansion of our civilization to a New World. As Zubrin observed in *The Case for Mars*: "Democracy in America and elsewhere in Western civilization needs a shot in the arm. That boost can only come from the example of a frontier people whose civilization incorporates the ethos that breathed the spirit into democracy in America in the first place. As Americans showed Europe in the last century, so in the next the Martians can show us the path away from oligarchy and stagnation." (Zubrin 303)

As a civilization, the road to Mars runs through a restoration of culture. This requires commitment on the part of individuals to break through the cults of personal peace and affluence. It requires wholehearted commitment to a goal larger than ourselves—not merely the opening of a New World, but the restoration of the Old. This commitment is not simply a matter of joining or contributing to the right organization; instead, the new Renaissance must begin *within ourselves*. The revival of culture begins with a change in the hearts and priorities of individual men and women.

In his book, *The Discoverers*, historian Daniel Boorstin noted, Two grand assumptions, which lay beneath talk of the Renaissance, shaped future thinking about man's role in all history. First, belief that every age somehow exuded a prevailing spirit—what German scholars called the Zeitgeist, what Carl Becker called the 'Climate of Opinion'—which favored certain notions and institutions. Second, that within these limits, men had the power to make history. Renaissance men made a Renaissance. (612)

What will be the spirit of our age? What *cultus* will shape our culture? To ignore the state of our civilization—or even, perversely, to revel in it—does not ameliorate one's responsibility. To fail to make a commitment to restoring our civilization is still to make a choice. The majority, perhaps, will continue to deny the gravity of our situation and choose what they perceive to be their personal advantage over the needs of others. But for many, the measure of that which has been lost is now being felt, and they are desirous of a restoration of that which was best in the age now passing.

The last time something like this happened was five or six hundred years ago, involving a small minority of people, which is not what is happening now. At that time men began to look back at the achievements and the letters and the art of Greece and Rome, idealizing them, and emulating them. (All art begins with emulation.) That was the Renaissance, a re-birth: the word is telling. It marked the beginning of modern historical consciousness—although that was imperfect and incomplete, because of its almost unrestricted idealization of the Classical Age, of the ancients. ... They took their inspiration from *two* ages away, farther back. This is not happening now. Something else is: our respect and admiration for the age that is now past but existed immediately before our times and which in many ways is still close to us and extant within us. (Lukacs 32–33)

The great strengths and weaknesses of the civilization which is passing were manifested within the hemisphere opened during the Renaissance and Reformation. The prospects for any new restoration of culture being a Red Planet Renaissance may rest with individuals who have already been born, who make a commitment in their own lives to live out a commitment to such a new Renaissance. That which has come before us is not lost inaccessibly; it remains for us to learn from its great strengths and weaknesses and seek out ways to apply such wisdom in the world around us—in our daily lives—and perhaps in the opening of a new frontier. If such a New World is opened to us and to our posterity, then the troubles of our age may seem as distant to our descendants as the troubles which exercised the courage and faith of our ancestors as they built the last New World.

# XI.
# A Shining City on a Higher Hill:
# Lessons from the Last Colonization of a 'New World.'

*For we must consider that we shall be as a City upon a Hill.*
*The eyes of all people are upon us...*
*—John Winthrop, 1630.*

## Introduction.

Nearly 400 years ago, on December 11, 1620, one hundred and one men, women and children landed at New Plymouth, an event which, as one historian describes it, was "the single most important formative event in early American history, which would ultimately have an important bearing on the crisis of the American Republic." ( Johnson 28) Of course, these settlers—a mix of Pilgrim Separatists and non-Pilgrim "Strangers"—were not the first colonizers of the 'New World'; indeed, they were not even the first English settlers. Two key factors distinguished these settlers from those who came before them: (1) their unshakable self-identity as a community, and (2) their commitment to a world view which shaped their motivation for colonization.

It is the purpose of this brief paper to set forth—albeit quite broadly—lessons which can be learned from the colonization of the 'New World' which could have much to teach us as we contemplate colonizing Mars. We will then examine the specifically religious

*This essay was presented at the founding convention of the International Mars Society, meeting in Boulder, Colorado in 1998.*

motivations which led to the landing at Plymouth which have reappeared in our age.

## Possible Motivations for Colonization of Mars.

For purposes of our study, we will consider three possible motivations for colonization of Mars: (1) military expansion or competition between colonizing nations, (2) economic exploitation of the natural resources of the colony, and (3) pursuit of religious freedom. Such an analysis does not discount the possibility of other motivations, or that individuals or nations might be influenced by more than one motivation. Still, it is our understanding that these were the most significant motivations.

### 1. Military Expansion or Competition.

Time does not permit a recounting of all of the struggles between the colonial powers in the New World; rather, our concern is with the early colonial division of the New World. This division first occurred between Spain and Portugal, but later included Sweden, England, France and the Netherlands.

Following Columbus' discovery of the New World, little time was lost in dividing up the 'spoils' between the European powers. Through a series of four papal bulls, Pope Alexander VI 'gave' the New World to Spain, setting down a line of demarcation running between the poles, "one hundred leagues towards the west and south from any of the islands commonly known as the Azores and Cape Verdes." (Boorstin 248) However, this action did not amuse King John II of Portugal, whose rather substantial navy placed him in a good bargaining position with Ferdinand and Isabella; this led to a more equitable line in 1494. (Boorstin 248–9)

However, although Spain and Portugal recognized this 'border', other Europeans did not. Indeed, one of the incidental 'fruit' of the Reformation was that it freed England from the ecclesio-political concerns connected with papal decrees. "In 1561, Queen Elizabeth I's Secretary of State, Sir William Cecil, carried out an investigation into the international law of the Atlantic, and firmly told the Spanish ambassador that the pope had no authority for his award." ( Johnson 9)*

Sweden and the Dutch Republic (having secured its independence from Spain in 1581) also established a 'presence' in North America in 1638 and 1624, respectively. (Nelson 4–5) Although the Swedish colonies were quite small (never numbering more than 400 inhabitants), the Dutch made no effort to encroach on their claims until 1655. "The reason for this toleration was that in the Thirty Years' War—which was then being fought in Europe, Sweden, and the Netherlands—Protestants were allied against Roman Catholic powers"†—a war which ended in 1648.

French Huguenot seamen automatically dismissed the Roman Catholic Church's claims and maintained that "the normal rules of peace and war were suspended beyond a certain imaginary line running down the mid-Atlantic. ... [T]he theory, and indeed the practice, of 'No Peace Beyond the Line' was a 16th-century fact of life." ( Johnson 9–10) Already in 1564–5, French Huguenots established a colony (Fort Caroline) on the northeastern coast of Florida. "Less than a month later, Florida being considered Spanish territory, the Spanish captain-general Peter Menendez attacked the settlement and massacred all the Huguenots. He was reported to have explained, 'I do this not as to Frenchmen but as to Lutherans [*Luteranos*].'" (Nelson 3) The Protestant Huguenots did not fair much better at the hands of the Portuguese, either. In 1555–6, Gaspard de Coligny, Admiral of France, sent two groups of colonists to settle in the harbor of Rio

---

* As early as 1530, the Lutheran Augsburg Confession declared, "If bishops have any power of the sword, that power they have, not as bishops, by the commission of the Gospel, but by human law, having received it of kings and emperors for the civil administration of what is theirs. This, however, is another office than the ministry of the Gospel." (*Concordia, or Book of Concord*, [St. Louis: Concordia Publishing House, 1922] p. 23.) Here we see a return to the earlier medieval church's distinction of the three estates of mankind: those who pray, those who fight, and those who work.

† In 1664, the Dutch, in turn, surrendered to English troops without firing a shot. (Nelson 6)

de Janeiro; many of the more than 300 colonists were chosen by John Calvin himself. However, the Portuguese struck the colony in 1560, hanging all of the colony's inhabitants. (Johnson 10) While one should not underestimate the religious nature of these attacks, it is impossible to dismiss the political motivations for both settlement and repression. International competition was no small motivation; the choice to settle (or destroy a settlement) expressed European tensions and conflicts.

In our present situation, however, a parallel motivation for colonization seems unlikely. The end of the U.S.-Soviet 'Cold War' eliminated much of the military competition motivation for colonization of Mars. This was not always the case, of course; it was the 'Cold War' which propelled the early 'space race', including the moon landings. (Lewis 5) (Still, the late Carl Sagan, among others, hoped that cooperation between the two nations might yield *exploration* of Mars, a crucial preliminary step to colonization [cited in Zubrin 279-282].) This does not mean it is impossible such a motivation could arise again in the future, although such a 'solution' seems worse than the 'problem' of motivating Mars colonization.

## 2. Economic Exploitation.

The second potential motivation, economic exploitation by a transient worker population, is possibly even more tenuous, since the success of such an approach rests on profits from hypothetical scientific advances or economically feasible exploitation and exportation of Martian natural resources, when relying on a population of workers motivated by a desire for a quick profit, not permanent colonization. Economic motivations require more than simply proving tremendous profits *could* be gained by going to Mars; it is necessary to convince multinational corporations and international banks that there is a guaranteed profit which equals (or exceeds) the profits of a comparably risky investment on Earth.* Such an endeavor will also be hampered by a lack of personal 'stake' in the development of Mars: For those drawn to Mars only by massive salaries, the new world will simply be a particularly unpleasant worksite with higher pay than usual—a place to be survived and *left as soon as possible*. The emphasis will be on stripping Mars of its resources, not developing a new civilization.

Again, the colonization of the 'New World'—particularly in the English colonies of Virginia—offer interesting historical lessons. First, consider the debacle of Sir Walter Raleigh's Roanoke colony. Certainly the settlement did not lack for resources: seven ships, over 100 men, troops led by experienced officers, an able scientist (Hariot) with instructions for scientific investigation, and knowledge of the local Indian language. Nevertheless, the assessment after the first year of the settlement was not encouraging: Ralph Lane (who commanded the settlement's troops) declared that the colonists "thought they would find treasure and 'after gold and silver were not to be found, as it was by them looked for, had little or no care for any other thing but to pamper their bellies.'" (Johnson 16) Nevertheless, 150 colonists sailed for Roanoke in 1587 and 114 (including sixteen women and ten children) remained behind when the fleet sailed for England. The results were, of course disastrous: the colony was apparently wiped out by Native Americans.

Roughly thirty years later (1607), Jamestown suffered nearly as disastrous a fate. Again, the problem was centered in colonists who were more interested in adventure (i.e., fast profits and a trip home) than colonization; "... lacking a family unit basis, the colony was fortunate to survive at all. Half died by the end of 1608, leaving a mere fifty-three emaciated survivors." (Johnson 25) John Smith's firm leadership (beginning a year after the establishment of the colony) slowed the colony's collapse, but he returned to England in 1609. Once again, of 400 new settlers arriving in 1609, by May 1610, "scarcely sixty settlers were still alive. All the food was eaten, there was

---

This is, of course, a marked weakness of this motivation in comparison to military competition, since nations rarely allow cost considerations to interfere with programs deemed necessary for national security, as can be amply proved from the arms build-up in the final years of the U.S.-Soviet 'Cold War'.

a suspicion of cannibalism, and the buildings were in ruins." ( Johnson 26) It took martial law under Thomas Dale to whip some order into the Virginia colony—and it took the establishment of *families* in the colony and a profitable export. In 1619, ninety unmarried women were brought to Virginia; "Any of the bachelor colonists could pur-chase one as a wife simply by paying her cost of transportation, set at 125 pounds of tobacco." ( Johnson 26) By 1616 Virginia tobacco had developed to the point where it was an exportable crop and in 1619 the first twenty African slaves arrived to grow tobacco.[*] The future course was set. Virginia survived in no small part because: (1) families were established in the colonies, moving away from disas-trous experiments with 'adventurers', (2) tobacco was developed to give the colony a valuable export, (3) indentured servants and slaves were imported to shore up the economy.[†] Certainly the English set-tlers did not 'invent' modern slavery,[a] but this development has had a profoundly harmful influence on our society to this day.

In his essay, "On Plantations" (1625), Sir Francis Bacon ex-amined the reasons for the failure of the earlier colonies. "He pointed out that any counting on quick profits was fatal, that there was a need

---

[*] Although austensibly only 'indentured' servants, it is doubtful any of this first group ever became free. ( Johnson 27)

[†] The increase in the number of slaves rose sharply after the 1670s: "The number of blacks in the South rose after the 1670s from 6 to 21 percent of the population, an influx concentrated in Maryland, Virginia, and above all in South Carolina, which in 1700 counted 43 percent of its people black. By that year over 80 percent of blacks in America lived in the South. Meanwhile, their numbers in the North remained fairly static—perhaps 7 percent of the population." (Hawke 128) Concerning the economics of slavery, see Hummel 38-52.

[a] Indeed, the Portuguese immeshed themselves in the economics of slavery a full forty years before Columbus' famous voyage. "The Portuguese entered the slave trade in the mid-15th century, took it over and, in the process, transformed it into something more impersonal, and horrible, than it had been either in antiquity or medieval Africa. The new Portuguese colony of Madeira became the center of a sugar industry, which soon made itself the largest supplier for western Europe. The first sugar-mill, worked by slaves, was erected in Madeira in 1452." (Johnson 5)

for expert personnel of all kinds, strongly motivated in their commitment to a long-term venture, and not least, that it was hopeless to try to win over the Indians with trifles 'instead of treating them justly and graciously.' ..." ( Johnson 18) All of this does not automatically doom the usefulness of short-term 'adventurers' in a future Martian colony—a certain number of them in certain specializations may be indispensable. But certainly the Virginia colony provides powerful warnings against relying heavily or exclusively on temporary workers; if nothing else, the cost of rotating hundreds of colonists 'home' every few years seems an unnecessary burden on the financial viability of a young Martian colony. We must keep in mind that initially Mars, like seventeenth century America, will not be an extremely wealthy place. As one historian observed:

> Even after the whole of the seaboard from stony Maine down to the swampy Florida frontier had been occupied by the British, the North American colonies would be of small value to Britain except as suppliers of tobacco, furs, naval stores, some dyestuffs and foodstuffs, and a few other products of no great consequence. Barbados or Jamaica alone—or, for that matter, some other islands in the West Indies—seemed worth more to London than all the thirteen mainland colonies put together.
>
> True, the North American colonies served in some degree to hold in check England's French and Spanish rivals in the New World, and therefore they were worth defending in time of crisis. In general, nevertheless, those North American territories were disappointing to the British trading companies that made the early settlements; they were disappointing to the great English proprietors—some, noblemen; others, commoners—who for decades controlled the vast empty lands; they were sufficiently disappointing even to the English kings who eventually asserted direct sovereignty over the Thirteen colonies. (Kirk 303)

### 3. Pursuit of Religious Freedom.

We turn, therefore, to the third motivation under consideration: religious freedom. The first two motivations sought to exploit the New World for advantage in the Old World. The motivation of religious freedom, however, is only incidentally interested in the concerns of the Old World; it is, if anything, a flight from the Old World.

It is our contention that religious freedom is the most fruitful motivation for colonization in terms of stability, steady growth, and cultural cohesion. The desire for religious freedom was the guiding motivation for successful English settlement in New England, efforts that were initially quite meager in terms of personnel and financial resources. The pursuit of religious freedom is a motivation powerful enough to move men and women to leave behind land, home, family—even a world—to build a new life.

It has been observed that: "Between the abortive Armada in 1588 and conclusion of the Thirty Years' War in 1648, Europe was a fascinating but fearful place. Transitional paroxysms of the later Renaissance and uncontrolled tremors of several sorts were felt everywhere in western Christendom. In England especially, for more than sixty years, there were profound crises of mind and spirit, reflected in religion, literature, and political thought." (Kammen 117) In 1588, the Spanish Armada attacked England with 130 ships and 30,000 men, confident that their forces (and a hoped for, but unrealized, Catholic uprising) would return Protestant England to the Roman Catholic fold. (Walker 524) Thirty years later, the Thirty Years' War began, and again Spanish troops (and those of other Roman Catholic territories) were on the move against Protestant forces, this time on the Continent. Indeed, the fall of the Bohemian forces in the Battle of White Mountain (November 8, 1620) was nearly simultaneous with the Pilgrims' landing. (Walker 530) Fear that European Protestantism would fall utterly to Roman Catholicism was a major concern motivating Puritan immigrants, particularly during the Great Migration of the 1630s. As Thomas Hooker declared in a 1630 sermon:

Will you have England destroyed? Will you put the aged to
trouble, and your young men to the sword? Will you have
your young women widows, and your virgins defiled? Will
you have your dear and tender little ones tossed upon the
pikes and dashed against the stones? Or will you have them
brought up in Popery, in idolatry, under a necessity of perish-
ing their souls forever, which is worst of all?

... God begins to ship away his Noahs, which proph-
esied and foretold that destruction was near; and God makes
account that New England shall be a refuge for his Noahs
and his Lots, a rock and a shelter for his righteous ones to
run unto; and those that were vexed to see the ungodly lives
of the people in this wicked land, shall there be safe.

(*The Puritans in America*, 68–9)

Even more than the foreign threat, however, Pilgrims (Separatist) and Puritans (non-Separatist)* were appalled by the decay of
English civilization and religion. The division within the English
church between Anglican and Puritan is well known: differences
over clerical attire and worship forms, the form of church government, communalism and discipline were all areas of disagreement.
(Kammen 125) However, the spread of immorality and the burden
of (perceived) overpopulation also motivated Puritans to emigrate
from England. Writing in 1629, John Winthrop gave the following
reasons for "the Intended Plantation in New England":

First, it will be a service to the church of great consequence to
carry the gospel into those parts of the world, to help on the
coming in of fullness of the Gentiles and to raise a bulwark
against the kingdom of Antichrist, which the Jesuits labor
to rear up on those parts.

2. All other churches of Europe are brought to
desolation, and our sins, for which the Lord begins already

* 'Separatist' in terms of consciously severing all ties with the state church of
England.

to frown upon us, do threaten us fearfully ...

3. This land [England] grows weary of her inhabitants, so as man who is the most precious of all creatures is here more vile and base than the earth we tread upon, and of less price among us than a horse or a sheep...

4. The whole earth is the Lord's garden and he hath given it to the sons of men, with a general condition, Genesis 1:28, *Increase and multiply, replenish the earth and subdue it*, which was again renewed to Noah. ...

5. We are grown in that height of intemperance in all excess of riot, as no man's estate almost will suffice to keep sail with his equals, and he who fails herein must live in scorn and contempt; hence it comes that all arts and trades are carried in that deceitful and unrighteous course, as it is almost impossible for a good and upright man to maintain his charge and live comfortably in any of them.

6. The fountains of learning and religion are so corrupted (as besides the unsupportable charge of education) most children (even the best wits and fairest hopes) are perverted, corrupted, and utterly overthrown by the multitude of evil examples of the licentious government of those seminaries...

7. What can be a better work and more honorable and worthy a Christian than to help raise and support a particular church while it is in the infancy?

8. If any such who are known to be godly and live in wealth and prosperity here shall forsake all this to join themselves to this church and to run a hazard with them of a hard and mean condition, it will be an example of great use both for removing the scandal of worldly and sinister respects which is cast upon the adventurers to give more life to the faith of God's people in their prayers for the plantation...

9. It appears to be a work of God for the good of his church in that he hath disposed the hearts of so many of his wise and faithful servants (both ministers and others) not

only to approve of the enterprise but to interest themselves in it... (*The Puritans in America*, 71–72)

The Pilgrims (and the Puritans who followed) were, therefore, different from those who came before: They did not come as individuals, but as a community. They did not come as adventurers, but as 'planters' (colonists). They came with a specific *vision* motivating their settlement—a revitalization of the Christian faith—and understood themselves bound up in a covenant with God in this task. Again, as Winthrop told his shipmates in 1630 on the way to the New World:

> Thus stands the cause between God and us. We are entered into covenant with him for this work. We have taken out a commission, the Lord hath given us leave to draw our own articles. We have professed to enterprise these actions, upon these and those ends, we have hereupon besought him of favor and blessing. Now if the Lord shall please to hear us, and bring us in peace to the place we desire, then hath he ratified this covenant and sealed our commission, [and] will expect a strict performance of the articles contained in it. But if we shall neglect the observation of these articles which are the ends we have propounded and, dissembling with our God, shall fall to embrace this present world and prosecute our carnal intentions, seeking great things for ourselves and our posterity, the Lord will surely break out in wrath against us, be revenged of such a perjured people, and make us know the price of the breach of such a covenant.
>
> Now the only way to avoid this shipwreck, and to provide for our posterity, is to follow the counsel of Micah, to do justly, to love mercy, to walk humbly with our God. For this end, we must be knit together in this work as one man. We must entertain each other in brotherly affection, we must be willing to abridge ourselves of our superfluities, for the supply of others' necessities. We must uphold

a familiar commerce together in all meekness, gentleness,
patience and liberality. We must delight in each other, make
others' conditions our own, rejoice together, mourn together,
labor and suffer together, always having before our eyes our
commission and community in the work, our community as
members of the same body. So shall we keep the unity of the
spirit in the bond of peace. The Lord will be our God, and
delight to dwell among us as his own people, and will com-
mand a blessing upon us in all our ways, so that we shall see
much more of his wisdom, power, goodness, and truth, than
formerly we have been acquainted with. We shall find that
the God of Israel is among us, when ten of us shall be able
to resist a thousand of our enemies; when he shall make us a
praise and glory that men shall say of succeeding plantations
[i.e., colonies], "the Lord make it like that of New England."
For we must consider that we shall be as a city upon a hill.
The eyes of all people are upon us, so that if we shall deal
falsely with our God in this work we have undertaken, and
so cause him to withdraw his present help from us, we shall
be made a story and a by-word through the world. We shall
open the mouths of enemies to speak evil of the ways of God,
and all professors for God's sake. We shall shame the faces
of many of God's worthy servants, and cause their prayers
to be turned into curses upon us till we be consumed out of
the good land whither we are agoing.

(*The Puritans in America* 90–91)

The Puritan decision to aggressively pursue a path of emigration was not accidental. Sir Robert Rich, Earl of Warwick, became
a member of the Virginia Company in 1612 and worked with other
like-minded Puritan gentry to reform England. However, should
that effort prove impossible,

[H]e wanted the alternative option of a reformed colony
in the Americas. Throughout the 1620s he was busy orga-

nizing groups of religious settlers, mainly from the West Country, East Anglia and Essex, and London—where strict Protestantism was strongest—to undertake the American adventure. In 1623 he encouraged a group of Dorset men and women to voyage to New England, landing at Cape Ann and eventually, in 1626, colonizing Naumkeag. John White, a Dorset clergyman who helped to organize the expedition, insisted that religion was the biggest single motive in getting people to hazard all on the adventure: 'The most eminent and desirable end of planting colonies is the propagation of Religion,' he wrote. ... He admitted: 'Necessity may oppress some: novelty draw on others: hopes of gain in time to come may prevail with a third sort: but that the most sincere and Godly part have the advancement of the Gospel for their main scope I am confident.' (Johnson 30)

## Conclusion—Religious Motivations for Emigration to Mars.

While our present study has focused on the influential Puritan immigrations, much could also be said regarding the influence religious freedom had on the formation of other colonies. Lord Baltimore's interest in chartering the Maryland colony was to provide refuge for fellow English Roman Catholics. (Walker 574) Pennsylvania was founded on the principle of religious freedom; William Penn was concerned with providing a refuge for fellow Quakers. (Walker 562, 577) In addition, much could be said concerning the later immigration of Christian communities into the colonies and the American Republic. Please indulge a few Lutheran examples: The Salzburg Lutherans were forced out of Austria because they would not convert to Roman Catholicism; these refugees eventually formed the Ebenezer colony in Georgia in 1734. (Finck 121) In 1843, a group of 1,600 Prussian Lutherans immigrated to New York and Wisconsin because they would not be part of a attempt by King Friedrich Wilhelm III to force a union of the Reformed and Lutheran churches. (Ewald 41–59) In 1839, 602 Saxon Lutherans arrived in St. Louis, having fled their homeland because of the rationalism and secularization they believed were destroying the church; the Lutheran Church—Missouri Synod they formed now numbers 2.5 million members in 6,000 congregations. (Camann 13–22)

Granted that religious freedom was a significant motivation for the successful colonization of the New World, is it possible this motivation would be operative in a future colonization of Mars? The stark reality is that Western Christians are probably experiencing a greater sense of cultural isolation and alienation than their seventeenth century forefathers. Then the division was between groups which still shared many points in common; even Deists still upheld the existence of God and were, in fact, outspoken defenders of the existence of natural law. Turner's groundbreaking study of the roots of Western unbelief, *Without God Without Creed*, contains the following conclusion concerning the present cultural division:

> Unbelief has transformed the hopes, aspirations, purposes, and behavior of millions of unbelievers. It has affected believers almost as remarkably. Wrangling over prayer in public schools is only one minor eruption, pointing to a major shift of the tectonic plates on which our culture moves. The option of godlessness has dis-integrated our common intellectual life, both in formal disciplines like philosophy, science, and literature and in those informal habits of mind by which we, as a culture, experience and order our world. God used to function as a central explanatory concept. As cause and purpose, the idea of God shaped and unified natural science, morality, social theories, psychology, political thought into one vaguely coherent (though very loosely assembled) approach to understanding humankind and the cosmos. ... [A]t the most fundamental level, God provided the frame of an agreed-upon universe in which to argue. Our web of shared assumptions has not unraveled altogether—without some unity, a culture collapses. But the traditional linchpin is

missing; our culture, in this sense, now lacks a center. (263–4)

The division between the Christian world view and that of the secular elite culture has grown increasingly distinct. In the words of Stephen Carter, "Our culture seems to take the position that believing deeply in the tenets of one's faith represents a kind of mystical irrationality." (Marshall 189) From a Christian perspective, many of the currents in Enlightenment, Modern, and Post-Modern thought have been as progressively de-humanizing as they have been increasingly agnostic or atheistic. Charles Darwin, Karl Marx and Sigmund Freud (and their intellectual heirs) are all understood to have simultaneously attempted to tear down God and those who were made in His image; both reason and faith become essentially meaningless terms when man's thoughts and actions are dictated by biology or economics. (Such determinism could be considered the agnostic's equivalent of John Calvin's double predestination.) For a Christian perspective, it has become an age in which, in the words of William Butler Yeats,

> Things fall apart; the centre cannot hold;
> Mere anarchy is loosed upon the world,
> The blood-dimmed tide is loosed, and everywhere
> The ceremony of innocence is drowned;
> The best lack all conviction, while the worst
> Are full of passionate intensity. (quoted in Kirk 7)

In summary, then, it is our contention that a significant number of Christians would embrace having 'somewhere to go.' As John Lewis observes in *Mining the Sky*, "It was the search for freedom of religion that brought most of our ancestors here, and it will be the search for freedom from religious, political, and ethnic persecution that will send the first colonists forth into space." (240) As with the settling of the New World, Christians colonists setting out for Mars would do so as a community seeking the freedom to worship the Triune God who created heaven and earth, redeemed us from our sin, and sanctifies us through His Word and Sacraments. This unity of faith will, Lord-willing, provide the motivation to endure the hardships of opening up a second new world, perhaps on the four hundredth anniversary of that historic landing at Plymouth.

Robert Cushman, numbered among the Pilgrims reaching our shores in November, 1620, wrote a defense of immigration to the 'New World' which was included in the work known as "Mourt's Relation" (1622), the first published account of the Pilgrims' arrival. His concluding words are so *apropos* that they will be our last word, as well.

> To conclude, without all partiality, the present consumption which groweth upon us here [in England], whilst the land groaneth under so many close-fisted and unmerciful men, being compared with the easiness, plainness and plentifulness in living in those remote places, may quickly persuade any man to a liking of this course, and to practise a removal, which being done by honest, godly and industrious men, they shall there be right heartily welcome, but for other of dissolute and profane life, theirs rooms are better than their companies. For if here, where the Gospel hath been so long and plentifully taught, they are yet frequent in such vices as the heathen would shame to speak of, what will they be when there is less restraint in word and deed? My only suit to all men is, that whether they live there or here, they would learn to use this world as they used it not, keeping faith and a good conscience, both with God and men, that when the day of account shall come, they may come forth as good and fruitful servants, and freely be received, and enter into the joy of their Master. (96)

# XII.
## Does Humanity have a Destiny on Mars?

When I was submitting my abstracts for this year's meeting of the Mars Society (and doing so, as usual, at the last moment), Maggie Zubrin posted me back to ask me whether I would consider doing *another* presentation, this time on the topic, "Why Mars?" I must admit I experienced a certain degree of trepidation at her request. "Why?" is often a delightfully dangerous question. "Why?" challenges simply taking things for granted, and that often makes people nervous. So, if you ask a theologian and church historian a question like this, you'd better be prepared for the reply. Perhaps not content to let "Why Mars?" be 'edgy' enough, I chose to rephrase the question as, "Does Humanity have a Destiny on Mars?" For some people, "destiny" is a sufficiently teleological term to immediately set off alarms. But does "destination" (a world which is closely related to "destiny" all the way back to their Latin 'roots') have the same impact? If the question were phrased, "Should Mars be a Destination for Humanity?," then I really would be 'preaching to the choir' here in Chicago! However, my 1972 edition of *Webster's New World Dictionary* gives the following definition for "destination": "1. [Rare] a destining or being destined, 2. the end for which something or someone is destined, 3. the place toward which someone or something is going or sent" (p. 383). It is interesting that common usage is often limited only to the third possible meaning for this word. Having a *destination* is inextricably caught up with having a *destiny*. What troubles many people is the notion that they are not the only one influencing or determining their *destination*.

*This essay was originally presented at the 2004 convention of the Mars Society, in Chicago, Illinois.*

"Why Mars?" The question is inextricably caught up in a link of *destination* and *purpose* (a term some might find more palatable than *destiny*). How one chooses to answer this question is inherently subjective, relying on the sense of purpose one brings to it. Most would probably rephrase the question as, "Why should Mars be a destination for humanity; what does one intend to achieve there?" Thus, my thoughts in this presentation center in the conjunction between human and divine *purpose*.

For the Christian, seeking to conform one's own will and purpose to the divine will is a central aspect of the Christian life. As Dr. Daniel Osmond observed:

> Our knowledge of God's Purposes is limited, but one thing is abundantly clear from the Judeo-Christian Scriptures and is authenticated by the best understanding of the historic Christian Church: God calls His people to Purposeful involvement in the world. Claiming divine authority, Jesus Christ commissioned His disciples to be engaged with Him wholeheartedly in praying for, and working out divine Purpose, "Your kingdom come, your will be done on earth as it is in heaven," and "Go and make disciples of all nations." ...
>
> Throughout Judeo-Christian history, strong belief in biblically inspired Purpose has translated into life-changing action. Such actions have been an outflow and manifestation of belief, not incidental to it. The value of the beliefs has been tested by the quantity and quality of actions they have produced and by their perseverance in the face of obstacles. It is one thing to do good things now and again: it is something else to initiate good things and to keep on doing them against the tide of popular opinion. ... There is a strong connection between knowing Purpose, glimpsing a vision of the "celestial city" and working hard to build a better world in the here and now. (Osmond 136–7)

As an acknowledgment of divine purpose has declined in certain circles, there was a concomitant decline in a sense of an overarching purpose to human activity. Sociologist Douglas Popora notes: "The problem in the modern or postmodern world is a pervasive loss of emotionally moving contact with a good that is ultimate, a contact that was once provided by the sacred. We are still emotionally moved by goods, but they tend to be goods that are less than ultimate—family, friends, and material possessions. As a consequence, the whole of our lives is without any overarching moral purpose." (71–2) Such a decline in an awareness of purpose, or its reduction to motivations of a lower order, was not necessitated by any advance in knowledge, but in the choice to adopt a different perception. In the words of Porpora: "It is not so much that metanarratives are not longer possible as that evidently we no longer seek them. Of course, we may also, as a consequence, no longer know who we are. The fragmentation of self that postmodernist philosophy takes to be a brute, ontological fact of our existence may, rather, be a result of a temporally specific disconnection from metanarratives." (132) Again, "If to know who we are is to know our place in the cosmos, then we cannot lose our place in the cosmos without losing ourselves as well." (152) People who have been indoctrinated to believe that their lives are essentially meaningless are likely to have very narrow horizons. They are certainly far less likely to sacrifice their own comfort and well-being for the good of others, or for the sake of future generations, or simply because "it's the right thing to do." A people without overarching purpose do not accomplish the advances that *matter* and *endure*.

I believe that a helpful answer to the question "Why Mars?" may be linked to the question concerning the purpose of life in the universe. During the past few decades, a significant number of Anthropic "Coincidences" have been identified which demonstrate than the laws of nature, our universe, and Earth in particular are unreasonably well-suited for life. This understanding, because it so directly opposes the philosophical schools which have purported the meaninglessness of existence, was initially advanced with some hesitation. Variations of the so-called Weak Anthropic Principle essentially argued, "Well, of course this is what inhabited universes look like. We're here, aren't we?," or declared, "It must not be too hard for life to arise, because we're here, and there's nothing special about Earth." With all due respect to the purveyors of such arguments, they really don't get to the point.

The literature on variations of stronger versions of the Anthropic Principle is becoming so vast and varied, the field cannot be given a proper overview in this presentation—which says a great deal about the vitality of this area of research. One work which I have found particularly helpful, and accessible, is Michael Denton's Nature's Destiny; the volume provides a summary of findings in many fields of science which contribute to the host of "coincidences." Please pardon a fairly lengthy quotation from the conclusion of Denton's work:

> We may not have final proof that the cosmos is *uniquely* fit for life as it exists on earth—because the possibility of alternative life cannot yet be entirely excluded—but there is no doubt that science has clearly shown that the cosmos is *supremely* fit for life as it exists on earth. For we have seen, the existence of life on earth depends on a very large number of astonishingly precise mutual adaptations in the physical and chemical properties of many of the key constituents of the cell: the fitness of water for carbon-based life, the mutual fitness of sunlight and life, the fitness of oxygen and oxidations as a source of energy for carbon-based life, the fitness of carbon dioxide for the excretion of the products of carbon oxidation, the fitness of bicarbonate as a buffer for biological systems, the fitness of the slow hydration of carbon dioxide, the fitness of the lipid bilayer as the boundary of the cell, the mutual fitness of DNA and proteins, and the perfect topological fit of the alpha helix of the protein with the large groove of the DNA. In nearly every case those constituents are the only available candidates for

their biological roles, and each appears superbly tailored to that particular end.

   If these various constituents—water, carbon dioxide, carbonic acid, the DNA helix, proteins, phosphates, sugars, lipids, the carbon atom, the oxygen atom, the transitional metal atoms and other metal atoms from groups 1 and 2 of the periodic table, sodium, potassium, calcium, and magnesium—did not possess precisely those chemical and physical properties they exhibit in an aqueous solution ranging in temperature from 0° C and about 75° C, self-replicating carbon-based chemical machines would be impossible. And it is not only microorganisms that the cosmic design has "foreseen." Many of the properties and characteristics of life's constituents seem to be specifically arranged for large, complex, multicellular organisms like ourselves. The coincidences do not stop at the cell but extend right on into higher forms of life. ...

   In short, science has revealed a *vast chain of coincidences which lead inexorably to life* on earth—not just microbial life but all life on earth, including large, air-breathing organisms like ourselves—a chain of adaptations which leads from the dimensions of galaxies, through the physical conditions in the center of stars to the heat capacity of water and the atom-manipulating capacities of proteins, and on eventually to our own species and our ability to comprehend the world. (381–2)

However, other scientists have demonstrated that although the universe may be 'unreasonably' well-suited for life, the existence of life (or at least intelligent life) is still vanishingly rare in the universe. Ward and Brownlee maintain what seems generally to be a highly credible scientific presentation in their book, *Rare Earth*, that advanced life is vanishingly rare in the universe: "We are one of many planets. But as we have tried to show in this book, perhaps not so many as we might hope—and perhaps not so many that we will ever, however long the history of our species, find *any* extraterrestrial animals among the stars surrounding our sun." (278)

Perhaps this begins to hint at the reason why the question, "Why Mars?," has really bothered me for some time, though perhaps not in the same way that others might address this question. At least as far back as '97 or '98, I found myself puzzling over variations of this question. For me, the question, "Why Mars?," is not so much an issue of economics, development, or 'progress' (whatever that term might mean in this context). For me, the question has taken on more the form of, "Why is there a *Mars*, anyway?" One can state the question in as basic a form as this; we have become accustomed to taking it for granted that a relatively habitable planet is "next door" to us. There is no known necessity that there would be any other rocky, terrestrial planets in the solar system; one could just as easily imagine a system with nothing but Earth and the gas giants. (Of course, it's hard to imagine Earth without the gas giants because Earth might well have been sterilized by cometary impacts.) Then again, Venus is hardly appealing, given surface conditions which present problems for human exploration (let alone colonization) which are essentially insurmountable in the foreseeable future. Mercury's proximity to the sun makes it nearly as uninviting as Venus, though it does enjoy a small advantage over Venus. The simple fact that Earth has a relatively accessible neighbor which has surface conditions which are not hopelessly intolerable for human exploration is simply astounding.

Many of the most salient benefits are well-known and easily summarized. First, Mars in near enough to the Sun to make it relatively easy to reach; the six month trip which is most readily cited as the likely length for the first voyages is already within human experience and, although problematic, it is not a "show-stopper." Mars is also near enough to the Sun that several apparently-sound terraforming schemes have been proposed—something which (arguably) *cannot* be said for Venus or Mercury. Mars is also large enough to hold an atmosphere (a marked advantage over the moon), and although it is *thin* (to say the least!), that is a far better circumstance than that found on the moon or Venus. Concerning Mars, one may speak of a "runaway greenhouse effect" (Zubrin 109) as a good thing. The utility of Mars's atmosphere to human exploration and colonization has been well-documented (most famously by Dr. Robert Zubrin).

In addition, the Martian rotation 'day' is almost the same length as Earth's, as is its axial tilt. There is an obvious benefit of such similarity of length of day for circadian rhythms (documented to exist in both animals and plants). Thus it has been observed, "*Mars and Earth are the only two locations in the solar system where humans will be able to grow crops for export.*" (Zubrin 1996, 222) If Mars had a ten hour or 45 hour day, the situation would obviously be quite different and could pose a fundamental problem for colonization. Mars's tilt on its axis is quite strikingly similar to Earth when one considers that Mars's equatorial inclination *has* been as much as 60 degrees in the past. Although the length of the seasons is far different on Mars, the Earth-like inclination poses less of a problem for settlement.

Just as importantly, Mars may offer the best possibility in the solar system of having the natural resources a colony would need. At this point, any definitive estimates will have to await the arrival of geologists. Nevertheless, this has not prevented some "educated speculation":

> Mars probably lacks ore deposits as diverse and rich as those on Earth's surface, although some ores probably formed during Mar's early wet era and during its longer-lasting volcanic episodes. So Mars probably comes closer than any other planetary body in the Solar System to matching the diversity of Earth's ore deposits. Indeed, iron may be more abundant in Mars's crust than Earth's, since Mars probably has not undergone as much internal differentiation as Earth. (Gonzalez and Richards 85)

One thing that is becoming increasingly clear is that Mars can meet the crucial need of supplying water. Although not as easily accessible as Earth, Mars certainly seems prepared to yield all the water one could reasonably need—something which cannot be said for Venus, the moon, or Mercury—and which, again, need not have been the case at all. It was not long ago, after all, that some of the 'best science' speculated whether Mars had much accessible water at all.

Certainly many points of unreasonable favorability could be offered. All of this seems to raise the question: but couldn't we imagine a Mars that would be *more* perfect? I expect Voltaire to pop in at any moment, sneering at anyone who would dare to claim we lived in the best possible world. I'm sure we could come up with a list of things we might imagine would make Mars 'better' (easier to study, easier to colonize, easier to terraform), but I end up wondering whether such a list would be founded in ignorance. Couldn't it be a little warmer, have a little more atmosphere, maybe running water? An Earth-like magnetic field would be nice, while we're at it, and a less elliptical orbit.* Yes, those are all reasonable questions, and we could quickly add more. The point of my presentation is not to describe Mars as necessarily the *perfect* 'starter kit' for an interplanetary civilization; rather, I am simply acknowledging that Mars is *unreasonably* well-suited for us. There is no known law of nature which would dictate that our solar system had to have a planet such as Mars.

Try to imagine the future of space exploration *without* Mars. Yes, man might go back to the moon, and might even eventually travel throughout the solar system. We might even build massive space stations or colonize the asteroids. But it *has* been a long time since the last flight to the moon, and we *still* haven't gone to Mars. It is hard to imagine humanity become interstellar without first becoming interplanetary. Keeping in mind Tsiolkovsky's oft-cited comment about humanity leaving the 'cradle,' it is worth remembering that a child must crawl, and then walk, before moving on to more adventur-

---

* Of course, with a less elliptical orbit, the study of astronomy would have been set back, since the Mars' orbit was helpful factor in studying the true structure of solar system.

ous things. The dream of such an expansion of human civilization is much more imaginable with Mars than without.

"Why Mars?" The answer is the cliché answer of the mountain climber—because it's there—which one begins to realize is an answer which is a lot more profound that it might sound at first. I am probably among the more vocal critics when it comes considering the kind of civilization which one should establishing on another world. But the simple fact that we can ask the question, "Why Mars?" is a profound wonder, if one truly stops to consider it.

It is a matter of faith whether one believes that life—especially human life—has purpose, meaning, and destiny. Of course, those who assert that life is meaningless are also asserting a philosophical (not scientific) position. But I do not believe that hope and purpose are illusions. Many people lost their way for a time, led into the paths of despair by a false (and ultimately tautological) line of reasoning. To be creatures of purpose means living up to what we were created to be; it implies stewardship and responsibility, and in a fairly crassly materialistic, self-centered age, there are many who indulge notions of meaninglessness and irresponsibility as a cop-out. There is, I suspect, some comfort in the knowledge that such individuals (and societies) are also not likely to spend their lives and fortunes on anything but themselves. They will not open a new world.

The answer to the question, "Why Mars?" is found in purpose—a sense of purpose in man which ultimately must come from above and extend beyond himself. Does humanity have a destiny on Mars? The answer to that question, apart from divine revelation, must await the passage of time. A Christian does not base his or her ultimate purpose in life in the things of this world—or Mars. We are to live out our lives in love of God and love of neighbor; and our daily failure in such conduct keep us mindful of the constant need for repentance and renewal in our walk. St. Peter wrote to the Church in his second Epistle:

> But the day of the Lord will come as a thief in the night, in
> which the heavens will pass away with a great noise, and the

elements will melt with fervent heat; both the earth and the works that are in it will be burned up. Therefore, since all these things will be dissolved, what manner of persons ought you to be in holy conduct and godliness, looking for and hastening the coming of the day of God, because of which the heavens will be dissolved, being on fire, and the elements will melt with fervent heat? Nevertheless we, according to His promise, look for new heavens and a new earth in which righteousness dwells. (3:10–13)

Secure in such a promise as that which Jesus has made to His people, "lo, I am with you always, even to the end of the age" (St. Matthew 28:20), the Christian lives as one who has purpose—purpose in the nearest neighbor or as far as the Lord shall lead him. For it is as David declares in Psalm 19: "The heavens declare the glory of God; and the firmament shows His handiwork." (v. 1)

# XIII.
## Science and Religion in the 'Space Age.'

Talking about the relationship between science and religion has been a good way either to empty a room or spark a fight. (So I'm happy to have all of you here tonight, and in a peaceful frame of mind.) Obviously, the conflict hasn't been limited to the academic circles of theologians and scientists. The struggle between science and religion is lived out in such mundane circles as school board meetings. It is experienced by every pastor as he teaches confirmation, and every confirmand who knows that God's Word teaches something very different from what may be taught at school. As a result, every one of us is challenged from a very early age to wrestle with this tension between science and religion, and people tend to react the most strongly when they are the least secure in their position.

Many Christians are not comfortable with such talk; there is a vague (or not so vague) sense in their minds that much of what the scientific community says is not consistent with what God's Word teaches, and they view science as a threat to that which is most important in their lives. Because our civilization enjoys tremendous material blessings as the fruit of scientific study, many Christians feel uncomfortable (or inadequate) challenging the claims of scientists, even when such scientists are speaking on topics which are not their expertise.

However, I know from personal experience that there are many scientists who are just as uncomfortable; they are dissatisfied

*This essay is adapted from a presentation at the Higher Things conference for Lutheran youth held at Seattle Pacific University, August 2004.*

with the sweepingly anti-religious claims of some within their community. As Astronomer Robert Jastrow observed, "When a scientist writes about God, his colleagues assume he is either over the hill or going bonkers." (9)

I believe than anything that evokes such strong emotions, is probably something that we need to talk about. The tensions between the scientific and the religious communities haven't gone away by ignoring them (or by ignoring each other). Just as certainly, peace has not been restored by trying to work out a "half a loaf" compromise, thinking that if both sides simply give up a few things, it will lead to peace. Both of these approaches have been tried before, to no avail. Professor John Haught observed that "there are at least four distinct ways in which science and religion can be related to each other" (*Science and Religion* 9). He identified these four ways as follows:

> 1) *Conflict*—the conviction that science and religion are fundamentally irreconcilable;

> 2) *Contrast*—the claim that there can be no genuine conflict since religion and science are each responding to radically different questions;

> 3) *Contact*—an approach that looks for dialogue, interaction, and possible "consonance" between science and religion, and especially for ways in which science shapes religious and theological understanding.

> 4) *Confirmation*—a somewhat quieter, but extremely important perspective that highlights the ways in which, at a deep level, religion supports and nourishes the entire scientific enterprise. (9)

These four categories do seem to cover most of the possible options, although I would add one which is a modification (but an important one) on the fourth category—confirmation. I believe it is important to reword it as follows according to the higher order of knowledge conveyed by God's Word: "Confirmation—an extremely important perspective that highlights the ways in which, at a deep level, the scientific enterprise confirms the teachings of Holy Scripture.''

Before proceeding further, perhaps a few autobiographical points are in order. My own interest in the relationship between science and religion extends back to childhood; I've been an avid science fiction fan since I was *quite* young (yes, I remember seeing *Star Wars* when it was first released), and watching the Apollo 11 landing on television is among my earliest memories. Computers and the heavens have been my particular areas of interest. While in high school, when I wasn't using my 48k home computer, I used the computers at the local community college (where, when writing FORTRAN programs, we entered code using punch cards), and a good modem connection to the university's computer was 300 baud (yes... 300 *bits* per second). I entered college as a computer science major, and loved the field—but it wasn't *enough*. And the Lord led me in a different way.

All the while, therefore, while studying theology, studying the sciences, and the *history* of science has continued to fascinate me. The intersection of these two interests came for me in 1997 and 1998, when I read Dr. Robert Zubrin's book, *The Case for Mars*. His 1996 book offered a viable, cost-effective approach to exploring Mars, the first true step (after the moon) in exploring the Solar System. In 1998, the Mars Society was established through the efforts of Zubrin and other scientists—some in the private sector, some working for NASA. I was a speaker at the founding convention, and I've spoken at several more of the annual conventions since then.

Major themes in my presentations to the Mars Society have included the relationship between the Christian faith and the renewal of civilization, space exploration, colonization, and the ethics of such activities as terraforming (the science of changing planetary environments to make them habitable for humans and other forms of life from Earth). Rather than meeting with hostility, I've found that most of the people who attend the conferences are quite open to hearing an explicitly Christian viewpoint on these topics. Over the years, I've had the opportunity at these conferences to meet many

Christians who work for NASA or within academia. Almost all of them have shared with me over and over again how *frustrating* it is that the Church has been frightened away from engaging in the discussion of scientific topics. Far too many have told me how family or even pastors denegrated science and scientific research as somehow 'ungodly' or even sinful. Many of those who've said such things have told me they came from an evangelical or charismatic background, but that has not been the case for all of them.

I am haunted by the memory of one woman who came up to me following my paper at the 1999 convention. I was engaged in a discussion with a small group, when she commented to me, while passing, "If I'd had a pastor like you, I never would have left the Church." I mention this *not* for the sake of any self-gratification; after all, I hadn't done or said anything that almost any confessional Lutheran pastor couldn't have said, after a little research. But the incident has haunted me because it tells me there are many pastors who *aren't* saying the right thing.

The Lord did not give us minds so that we would ignore them. Of course, a person is not more virtuous on account of greater knowledge; but neither is it a virtue to choose to remain ignorant. The scientific vocations are neither more, nor less, inherently Christian than any other. Those whom the Lord has given the aptitude and inclination for scientific pursuits should explore such a calling.

The first human activity witnessed in the Bible was, in a sense, scientific. We read in Genesis 2: "Out of the ground the LORD God formed every beast of the field and every bird of the air, and brought them to Adam to see what he would call them. And whatever Adam called each living creature, that was its name. So Adam gave names to all cattle, to the birds of the air, and to every beast of the field. But for Adam there was not found a helper comparable to him." (v. 19–20 NKJV) This 'giving of names' is called "taxonomy," which *Webster's New World Dictionary* defines as: "1. the science of classification; laws and principles covering the classifying of objects; 2. *Biol.* a system of arranging animals and plants into natural, related groups based on some factor common to each, as structure, embryology, biochemistry, etc. ..." To behold Adam before the Fall, engaged in a scientific endeavor, teaches us that the scientific mind is also part of the good creation.

However, this does *not* mean that everything which is proclaimed in the name of 'science' (in the broadest sense of that term) is true. Just as sin corrupts the *emotions* and *desires* of man, so, too, it corrupts the faculty of *reason*. Human history is littered with false doctrines, vain philosophies, and fallacious world views. And thus God's Word warns us in Colossians 2: "Beware lest anyone cheat you through philosophy and empty deceit, according to the tradition of men, according to the basic principles of the world, and not according to Christ." (v. 8) And St. Paul warned the Romans specifically of the damage which sin has done to man's recognition of the glory of God in the created order: "For since the creation of the world His invisible attributes are clearly seen, being understood by the things that are made, even His eternal power and Godhead, so that they are without excuse, because, although they knew God, they did not glorify Him as God, nor were thankful, but became futile in their thoughts, and their foolish hearts were darkened." (1:20–21)

As the conflict in the relationship between science and religion has its roots in the Renaissance, I believe we would profit from a brief look at three men whom I believe typify three different understandings of this relationship: Giordano Bruno (1548–1600), Galileo Galilei (1564–1642) and Johannes Kepler (1571–1630).

Giordano Bruno is best remembered today for his death: he was burned by the Roman Catholic Inquisition in 1600. Bruno and Galileo are often mentioned as the two scientists persecuted by the Inquisition, but Bruno's status as a "scientist" is dubious, at best. Bruno is of interest to us because of his views regarding the relationship between religion and science. As one scholar has noted:

> Bruno is chiefly celebrated in histories of thought and of science, not only for his acceptance of the Copernican theory [that the planets revolve around the sun], but still

more for his wonderful leap of the imagination by which he attached the idea of the infinity of the universe to his Copernicanism, an extension of the theory which had not been taught by Copernicus himself. ..

Yet is has also been observed that "Bruno's world-view is vitalistic, magical; his planets are animated beings that move freely through space of their own accord like those of Plato and Patrizi. Bruno's is not a modern mind by any means." (Yates 244)

Bruno was obsessed with magic and occultism; he desired—and worked for—a revival of what he thought was an ancient Egyptian form of magic and mysticism. He exploited the Copernican discoveries (if, in fact, he really understood them at all) to serve his world view. Again, in the words of Frances Yates:

Bruno's philosophy cannot be separated from his religion. It *was* his religion, the "religion of the world", which he saw in this expanded form of the infinite universe and the innumerable worlds as an expanded gnosis [or hidden knowledge] ... Copernicanism was a symbol of the new revelation, which was to mean a return to the natural religion of the Egyptians, and its magic, within a framework which he so strangely thought could be a Catholic framework. (355)

Bruno serves as a 'type' of all those who misuse the findings of science to push anti-Christian world views. Regrettably, there is no lack of such individuals, many of whom attempt to push "New Age" religious views under the guise of "science." He also serves as a 'type' of many of the aggressively-atheist evolutionists, who twists the data (consciously or not) to push an anti-religious agenda.

Galileo's situation was rather different from that of Bruno. Galileo was certainly among the most accomplished scientists of his age. He utilized the telescope (then a recent invention) to draw detailed images of the moon, observed the moons of Jupiter, and even the rings of Saturn (although the actual nature of the rings eluded him). But Galileo made a crucial error: he forgot the influence of sin on human reason. In his book, *Galileo's Mistake*, Wade Rowland observes as follows:

> Nature is also the revelation of God, he [Galileo] continues, but *nature is its own interpreter*, and never errs. In his words: "But nature, on the other hand, is inexorable and immutable; she never transgresses the laws imposed on her, or cares a whit whether her abstruse reason and methods of operation are understandable to men. For this reason it appears that nothing physical which sense-experience sets before our eyes, or which necessary demonstrations prove to us, ought to be called in question, much less condemned, upon the testimony of Biblical passages..." He concludes that certainties arrived at in physics, therefore, ought to be used to help interpret Scripture. But the reverse is not true.
>
> Here, in concise form, is what I have characterized as "Galileo's mistake." It is an error that has been understood by philosophers from the eighteenth century onward, from David Hume to Immanuel Kant to Thomas Kuhn, with increasing clarity. The mistake is in the belief that nature is its own interpreter. It is not. Nor is it the case, as Galileo claimed, that "it is not in the power of any created being to make things true or false." It is simply not correct to assert, as Galileo did, that there is a single and unique explanation to natural phenomena, which may be understood through observation and reason, and which makes all other explanations wrong. (137)

Living amidst the doctrinal contentions of the sixteenth and seventeenth centuries, Galileo knew that men could err in their interpretation of the Word of God (after all, he knew that mutually contradictory interpretations could not both be right), but his reasoning was flawed in imagining that science *could* deliver inerrant data and an inerrant interpretation of that data. Galileo thus serves as a type of all those who seek to reconcile science and religion, but to do so on what they consider to be the scientific "facts." There are *many* people who do this today. They assume that because 'science' tells us that the universe is fifteen billion years old, or that man evolved over millions of years from single celled organisms, that such an interpretation of the data of nature is *true*, and that the philosophical conclusions which are sometimes asserted from such claims are *true*, and that biblical accounts which contradict such claims must be reinterpreted to fit the claims of 'science.' There are many people who believe in a 'theistic evolution,' imagining that evolution occurred, and somehow God guided that process. (There are many problems with such theories, including the inability to explain the existence of death before sin, which all evolutionary theory necessitates.) Many of the Christians within the scientific community live with such tensions in their world view. Caught up in 'Galileo's mistake' is the tension between the providence of God and the mechanistic universe of the materialist who believes everything happens out of necessity.

Our third example is Johannes Kepler, who trained at the university to become a Lutheran pastor. Instead of becoming a pastor, Kepler became a mathematics teacher, and later an astronomer, serving as the astronomer to the emperor of the Holy Roman Empire. Kepler's world view was different from Bruno's and Galileo's in some very important ways, for he believed that God created the universe in such a way that He intended man to study it. As Kepler repeatedly observed in his book, *Epitome of Copernican Astronomy*, that Earth is the "home of the speculative creature": "For the earth was going to be the home of the speculative creature, and for his sake the universe and world have been made." (33)

When Kepler turned his thoughts to the heavens, he did so both as a theologian *and* as a scientist. He expected to find order—harmony—in the heavens because they had been created by the Triune God. This faith in an orderly universe is among the underlying assumptions of the entire scientific endeavor. It is rooted in a fundamentally Christian world view, and apart from the faith, there is no reason for presuming that the universe is orderly or comprehensible. Albert Einstein once observed, "science can only be created by those who are thoroughly imbued with the aspiration toward truth and understanding. This source of feeling, however, springs from the sphere of religion. ... I cannot conceive of a genuine scientist without that profound faith." (quoted in *Einstein and Religion* 94) In their recent book, *The Privileged Planet*, Gonzalez and Richards note:

> One of modern's science's most important biblical inheritances is the notion that lineal time is fundamental to the physical universe rather than an illusion. In other words, that cosmic history actually goes somewhere rather than merely in circles. In this way it distinguished itself from the cyclical view of time held by the Greeks. Also important was the biblical distinction between Creator and creation...
>
> Another significant biblical contribution is that since God was free in creating the world, nature is *contingent*. It might not have existed, or it might have had different properties from the ones it has. As a result, nature's properties must be discovered rather than merely deduced from the principles of logic or mathematics. ...
>
> Finally, since they believed that God is one and that human beings are created in God's image, medieval Christians and Jews could expect nature to have a sort of unity (to be a *universe*) and to be accessible to the human mind. These ideas, brought to fruition by interaction with the Greeks, were the seedbed from which natural science slowly grew. It's hardly a coincidence that science emerged in the time and place where these many factors converged. Although they are now forgotten, modern science draws on the interest of specific theological convictions. (228–9)

The simple fact that human beings than think *abstractly*, and that such abstractions render meaningful information about the world, is utterly inexplicable to the evolutionist or materialist (that is, someone who believes that there is no spiritual realm). Evolutionary theory might *attempt* to understand man's ability to procure food and reproduce; it does not explain how or *why* man has a mind capable of understanding quantum physics or of composing a sonnet. The materialist cannot explain why the higher mathematics conceived by the human mind have any correspondence to the way the world actually is. The evolutionist or materialist cannot explain why truth is beautiful to the human mind; why the common experience of scientists is that of knowing a theory is wrong because it lacks 'elegance.' Thus the dogmatic materialist has "come along for the ride"; his world view could not have given rise to modern science, but now he exploits 'science' (as Bruno did) to wage war against the very world view which made science possible.

God's Word clearly describes individuals such as the modern dogmatic materialist: "For since the creation of the world His invisible attributes are clearly seen, being understood by the things that are made, even His eternal power and Godhead, so that they are without excuse, because, although they knew God, they did not glorify Him as God, nor were thankful, but became futile in their thoughts, and their foolish hearts were darkened." (Romans 1:20–21 NKJV) The supposed-science they rely on to argue against the existing of a divine Creator is predicated on belief in the God revealed in Sacred Scripture. As Stephen Barr observes in his book, *Modern Physics and Ancient Faith*:

> What is most puzzling to the religious person about this materialist dogmatism is its lack of foundation. The religious dogmatist, after all, accepts certain truths as dogmas only because he believes them to have been revealed by God. But the materialist obviously cannot claim divine authority for his statement that only matter exists. ...
>
> ... *Nothing* is allowed by him to be beyond explanation in terms of matter and the mathematical laws that it obeys. If, therefore, he comes across some phenomenon that

is hard to account for in materialist terms, he often ends up by denying its very existence. ... The materialist lives in a very small world, intellectually speaking. It is a universe of huge physical dimensions, but very narrow for all that. There is no purpose in this universe. Even human acts are entirely determined by physical processes. (16–17)

Even those who believe in an 'old universe' (e.g., the 'Big Bang' theory) now acknowledge that the universe had a *beginning* and will have an *end*—contrary to all of the expectations of the materialists, who once believed that matter was eternal, without beginning or end. The study of the heavens and the earth presses the centrals question more now than ever: why is there *something* rather than *nothing*; and, what, then, is the *purpose* of it all? For those who are not willfully blind, the very heavens continue to proclaim the glory of their Creator. In the words of Psalm 19:

1. The heavens declare the glory of God; and the firmament shows His handiwork.
2. Day unto day utters speech, and night unto night reveals knowledge.
3. There is no speech nor language where their voice is not heard.
4. Their line has gone out through all the earth, and their words to the end of the world. In them He has set a tabernacle for the sun,
5. which is like a bridegroom coming out of his chamber, and rejoices like a strong man to run its race.
6. Its rising is from one end of heaven, and its circuit to the other end; and there is nothing hidden from its heat.

The psalmist acknowledges the universal witness of the heavens to God's glory; it is manifest to human intellect, based on natural observation, that there is a Creator. The observation of the motions of the heavens is part of the evidence for creation, for they speak to an order which does not arise of itself. Apart from the influence of original sin, the honest study of the creation (and especially the heavens) leads to knowledge of the Creator: "Day unto day utters speech, and night unto night reveals knowledge." All men even have some knowledge of the divine Law written on their hearts; in the words of Romans 2: "for when Gentiles, who do not have the law, by nature do the things in the law, these, although not having the law, are a law to themselves, who show the work of the law written in their hearts, their conscience also bearing witness, and between themselves their thoughts accusing or else excusing them" (v. 14–15).

But for as much as might be learned from the natural revelation—the testimony in creation to its Creator—such natural revelation tells us *nothing* of salvation. For this we need God's special revelation in Holy Scripture. thus the words of Psalm 19 continue:

> 7. The law of the LORD is perfect, converting the soul; the testimony of the LORD is sure, making wise the simple;
> 8. The statutes of the LORD are right, rejoicing the heart; the commandment of the LORD is pure, enlightening the eyes;
> 9. The fear of the LORD is clean, enduring forever; the judgments of the LORD are true and righteous altogether.
> 10. More to be desired are they than gold, yea, than much fine gold; sweeter also than honey and the honeycomb.
> 11. Moreover by them Your servant is warned, and in keeping them there is great reward.

The Law of God may be known to a limited extent from the study of nature; the divine Law is fully revealed in Holy Scripture. But the knowledge of the Law brings sorrow, too, for it reveals that we are sinners. How wonderful, then, that the means of salvation is revealed in God's Word! Thus the psalmist concludes:

> 12. Who can understand his errors? Cleanse me from secret faults.
> 13. Keep back Your servant also from presumptuous sins; let them not have dominion over me. Then I shall be blameless,

and I shall be innocent of great transgression.

14. Let the words of my mouth and the meditation of my heart be acceptable in Your sight, O LORD, my strength and my Redeemer.

The Psalm which begins by affirming the knowledge of God which may be gained from nature, concludes with that faith in the Lord, our *strength* and *Redeemer*, which is revealed only in Sacred Scripture. Guided by the truth of God's holy Word—not only the framework it provides us for studying the material world, but also the *reason* and *purpose* for creation which it clearly reveals—the Christian is the ideal scientist. It is a vocation not easily followed, for there are many within the scientific community who are in service to the inhuman ideology of materialism. In the face of often-overwhelming opposition and pressure to conform to a materialistic ideology, a Christian has to be firmly rooted in his faith if he is to make a true contribution to the sciences. However, for those who, by the grace of God, stand steadfast in their convictions, there is a crucial framework provided in understanding the *why* behind what *is*. As Robert Jastrow concludes in his book, *God and the Astronomers*:

> The scientist's pursuit of the past ends in the moment of creation.
>
> This is an exceedingly strange development, unexpected by all but the theologians. They have always accepted the word of the Bible: In the beginning God created heaven and earth. To which St. Augustine added, "Who can understand this mystery or explain it to others?" ...
>
> ... It is not a matter of another year, another decade of work, another measurement, or another theory; at this moment it seems as though science will never be able to raise the curtain on the mystery of creation. For the scientist who has lived by his faith in the power of reason, the story ends like a bad dream. He has scaled the mountains of ignorance; he is about to conquer the highest peak; as he pulls himself

over the final rock, he is greeted by a band of theologians
who have been sitting there for centuries. (106–7)

# Bibliography

Aristotle, *The Politics* (Oxford University Press, 1945)

Artigas, Mariano, *The Mind of the Universe—Understanding Science and Religion* (Templeton Foundation Press: Philadelphia and London, 2000)

Barr, Stephen M., *Modern Physics and Ancient Faith* (University of Notre Dame Press: Notre Dame, 2004)

Barzun, Jacques, *From Dawn to Decadence* (HarperCollins: New York, 2000)

Behe, Michael, *Darwin's Black Box* (Touchstone: New York, 1996)

Benford, Gregory, *The Sunborn* (Aspect: New York, 2005)

Bergaust, Erik, *Wernher von Braun* (National Space Institute: Washington, D.C., 1976)

Bergman, Jerry, *Wernher von Braun: The Father of Modern Space Flight—A Christian and a Creationist*, www.adam.com.au/bstett/BVonBraun96.htm (referenced August 9, 2005)

Bloom, Allan, *The Closing of the American Mind* (Simon and Schuster: New York, 1987)

Boorstin, Daniel J., *The Discoverers* (Vintage Books: New York, 1985)

Bova, Ben, *Faint Echoes, Distant Stars* (Perennial: New York, 2005)

Cahill, Thomas, *How the Irish Saved Civilization* (Doubleday: New York, 1995)

Camann, Eugene W., "The Saxon Migration to Missouri," in *Confessional Lutheran Migration to America*, (Eastern District, LC—MS, 1988)

Codevilla, Angelo M., *The Character of Nations*, (Basic Books: New York, 1997)

Connor, James A., *Kepler's Witch* (HarperCollins: New York, 2004)

*Cosmos, Bios, Theos—Scientists Reflect on Science, God, and the Origins of the Universe, Life, and Homo sapiens*, ed. by Henry Margenau and Roy Abraham Varghese, (Open Court: Le Salle, Illinois, 1992)

Coyne, George V., S. J., "Extraterrestrial Life and our World View," in *Many Worlds—The New Universe, Extraterrestrial Life & the Theological Implications*, ed. by Steven Dick (Templeton Foundation Press: Philadelphia and London, 2000), p. 177–188.

Cushman, Robert, *Mourt's Relation, A Journal of the Pilgrims at Plymouth*, intro. by Dwight B. Heath, (Applewood Books: Bedford, Massachusetts, 1963)

Davies, Paul, "E.T. and God," *The Atlantic Monthly* (September 2003) http://www.theatlantic.com/issues/2003/09/davies.htm

de León-Jones, Karen Silvia, *Giordano Bruno & the Kabbalah* (University of Nebraska Press: Lincoln, 1997)

Dembski, William A., *Intelligent Design—The Bridge Between Science & Theology* (InterVarsity Press: Downer's Grove, IL, 1999)

Denton, Michael J., *Nature's Destiny—How the Laws of Biology Reveal Purpose in the Universe* (Free Press: New York, 1998)

Diamond, Jered, *Collapse* (Viking: New York, 2005)

Dick, Steven J., *Plurality of Worlds* (Cambridge University Press: Cambridge, 1982)

Duhem, Pierre, *Medieval Cosmology—Theories of Infinity, Place, Time, Void, and the Plurality of Worlds*, ed. and trans. by Roger Ariew (University of Chicago Press: Chicago, 1985)

Ewald, Alfred H., "From a German Jail" in *Church Roots*, ed. by Charles P. Lutz, (Augsburg Publishing House: Minneapolis, 1985)

Finck, William J., *Lutheran Landmarks and Pioneers in America* (The United Lutheran Publication House: Philadelphia, 1913)

Fregosi, Paul, *Jihad* (Prometheus Books: Amherst, New York, 1998)

Gimpel, Jean, *The Medieval Machine* (Barnes & Noble: New York, 2003)

Gingerich, Owen, "Dare a Scientist Believe in Design?," in *Evidence of Purpose—Scientists Discover the Creator*, ed. by John Marks Templeton, (Continuum: New York, 1994)

Gingerich, Owen, *The Book Nobody Read* (Walker & Co.: New York, 2004)

156

Goldsmith, Donald, The Hunt for Life on Mars (Dutton: New York, 1997)

Gonzalez, Guillermo and Jay W. Richards, The Privileged Planet (Regnery: Washington, 2004)

Gorringe, Tim, A Theology of the Built Environment (Cambridge University Press: Cambridge, 2002)

Haught, John F., Science and Religion: From Conflict to Conversation (Paulist Press: Mahwah, NJ, 1995)

Haught, John, "Theology after Contact—Religion and Extraterrestrial Intelligent Life," in Annals of the New York Academy of Sciences 950:296-308 (2001)

Hawke, David Freeman, Everyday Life in Early America (Harper & Row: New York, 1989)

Heiser, James D., "Pilgrims Redux—Will Religious Communities Be Involved in Space Colonization?," Ad Astra, November/December 1998 (10:6)

Hughes, Thomas P., Human-Built World (University of Chicago Press: Chicago and London, 2004)

Hummel, Jeffrey Rogers, Emancipating Slaves, Enslaving Free Men (Open Court: Chicago and La Salle, 1996)

Huntington, Samuel P., The Clash of Civilizations and the Remaking of World Order (Touchstone: New York, 1997)

Jaki, Stanley P., "The Limits of a Limitless Science," in The Limits of a Limitless Science (ISI: Wilmington, 2000)

Jastrow, Robert, God and the Astronomers (W.W. Norton and Company: New York, 2000)

Paul Johnson, A History of the American People (Harper Collins: New York, 1998)

Kammen, Michael, People of Paradox (Cornell University Press: Ithaca, New York, 1980)

Kaplan, Robert D., The Coming Anarchy: Shattering the Dreams of the Post Cold War (Vintage Books: New York, 2000)

Kargel, James, Mars—A Warmer Wetter Planet (Springer-Verlag: Berlin, 2004)

Kepler, Johannes, *Epitome of Copernican Astronomy* (Prometheus Books: Amherst, 1995)

Kirk, Russell, "Civilization without Religion?," and "The Conservative Purpose of a Liberal Education," in *Redeeming the Time* (Intercollegiate Studies Institute: Wilmington, 1998)

Kirk, Russell, *The Roots of American Order* (Regnery Gateway: Washington, 1991)

Lewis, C. S., "Religion and Rocketry," in *The World's Last Night and other Essays* (Harvest: New York, 1960)

Lewis, John, *Mining the Sky* (Helix Books: Reading, Massachusetts, 1996)

Lukacs, John, *At the End of an Age* (Yale University Press: New Haven and New York, 2002)

Malove, Eugene, *The Quickening Universe* (St. Martin's Press: New York, 1987)

Manske, Charles L. and Daniel N. Harmelink, *World Religions Today* (Institute of World Religions: Irvine, CA, 1996)

Marshall, Paul, *Their Blood Cries Out—The Worldwide Tragedy of Modern Christians Who Are Dying for Their Faith* (Word Publishing: Dallas, 1997)

Melanchthon, Philip, *Orations on Philosophy and Education*, ed. by Sachiko Kusukawa (Cambridge: New York, 1999)

Murphy, Nancey and George F. R. Ellis, *On the Moral Nature of the Universe* (Fortress Press: Minneapolis, 1996)

Nauert, Charles G., *Humanism and the Culture of Renaissance Europe* (Cambridge: Cambridge, 1999)

Nelson, E. Clifford, *The Lutherans in North America* (Fortress Press: Philadelphia, 1980)

Osmond, Daniel H., "A Physiologist Looks at Purpose and Meaning in Life," in *Evidence of Purpose*, ed. by John Marks Templeton (Continuum: New York, 1994)

Porpora, Douglas, *Landscapes of the Soul, the Loss of Moral Meaning in American Life* (Oxford: Oxford, 2001)

*The Puritans in America, A Narrative Anthology*, ed. by Alan Heimert and

Andrew Delbanco (Harvard University Press: Cambridge, 1985)

Turner, *Without God without Creed* (The John Hopkins University Press: Baltimore and London, 1985)

Rowland, Wade, *Galileo's Mistake—A New Look at the Epic Confrontation between Galileo and the Church* (Arcade Publishing: New York, 2001)

Schaeffer, Francis, *How Should We Then Live?* (Crossway Books: Wheaton, IL, 1976)

Schäfer, Lothar, *In Search of Divine Reality—Science as a Source of Inspiration* (University of Arkansas Press: Fayetteville, 1997)

Sheehan, William & Stephen James O'Meara, *Mars—The Lure of the Red Planet*, (Prometheus Books: Amherst, NY, 2001)

Smith, Huston, *Why Religion Matters* (HarperCollins: New York, 2001)

Walker, Williston , et al., *A History of the Christian Church* (Charles Scribner's Sons: New York, 1985)

Walter, Malcolm, *The Search for Life on Mars* (Perseus Books: Cambridge, 1999)

Ward, Bob, *Dr. Space—The life of Wernher von Braun* (Naval Institute Press: Annapolis, MD, 2005)

Ward, Peter D., and Donald Brownlee, *Rare Earth* (Copernicus: New York, 2000)

Webster, Charles, *From Paracelsus to Newton—Magic and the Making of Modern Science* (Dover: Mineola, New York, 2005)

Wilker, Benjamin D., "Alien Ideas—Christianity and the Search for Extraterrestrial Life," *Crisis* (Nov. 4, 2002)

Wilkins, John, *The Discovery of a World in the Moon* (London, 1638)

Yates, Frances, Giordano Bruno and the Hermetic Tradition (University of Chicago Press: Chicago, 1964)

Zachmann, Randall C., "The Universe as the Living Image of God: Calvin's Doctrine of Creation Reconsidered," by *Concordia Theological Quarterly*, October 1997 (61:4)

Zubrin, Robert, *Entering Space* (Jeremy P Tarcher/Putnam: New York, 1999)

Zubrin, Robert, *The Case for Mars*, (The Free Press: New York, 1996)

# Other Titles Available from Repristination Press

*The Book of Concord*, hardcover, 360 p., $27.95—ISBN 1891469290

David Chytraeus, *Chytraeus on Sacrifice*, paperback, 152 p., $12.95— ISBN 1891469231

David Chytraeus, *A Summary of the Christian Faith*, paperback, 152 p. $14.99—ISBN 1891469444

Erdmann Fischer, *The Life of John Gerhard* (1723), hardcover, 464 p., $30.00—ISBN 1891469341

Johann Gerhard, *An Explanation of the History of the Suffering and Death of our Lord Jesus Christ (1622)*, hardcover, 330 p., $20.00— ISBN 1891469223

Johann Gerhard, *Daily Exercise of Piety (1629)*, hardcover, 96 p., $12.95—ISBN 1891469185

Johann Gerhard, *Sacred Meditations*, paperback, 302 p., $13.95— ISBN 1891469193

Paul R. Harris, *Why is Feminism so Hard to Resist?*, paperback, 166 pages, $12.95

Nicolaus Hunnius, *Diaskepsis Theologica (1626)*, hardcover, 543 p., $40.00—ISBN 1891469258

J.A. Quenstedt, *The Church*, hardcover, 154 p., $12.00—ISBN 1891469266

All prices are in U.S. dollars. Please include $2.00 shipping for each title, for delivery to U.S. address; please include $4.00 shipping per title for address outside the United States.

Send payment by check or money order for U.S. dollars to:

REPRISTINATION PRESS

P.O. BOX 173

BYNUM, TX 76630

## About the Author:

The Rt. Rev. Heiser earned his B.A. in Political Science from George Washington University, Washington, D.C. Beginning in 1987, he served for several years as a Research Associate at the National Center for Public Policy Research (D.C.), and as a Media Analyst for the Media Research Center (Alexandria, VA).

In 1995, Heiser earned his M.Div. from Concordia Theological Seminary, Ft. Wayne, Indiana, serving as the Graduate Assistant for Systematic Theology during the 1995–96 academic year.

In 1996 through 1997, Heiser served as the ordained Deacon of Our Redeemer Lutheran Church (Forsyth, IL), and as an Assistant Professor of Luther Bible College (Rockford, IL), teaching "The Gospel of Luke," and "Introduction to Apologetics."

Since 1998, Heiser has served as the Pastor of Salem Lutheran Church in Malone, Texas, while maintain his responsibilities as publisher of Repristination Press, which he established in 1993. In 2005, he was elected to serve as Dean of Missions for The Augustana Ministerium and in 2006 was called to serve as Superintendent/ Bishop of the Evangelical Lutheran Diocese of North America. He also serves as the current President of the Center for the Study of Lutheran Orthodoxy.

Heiser's publications include The Office of the Ministry in N. Hunnius' Epitome Credendorum (1996), as well as dozens of journal articles and reviews. He resides in Hillsboro, Texas with his wife, Denise, and children, John and Anastasia.

www.ingramcontent.com/pod-product-compliance
Lightning Source LLC
Chambersburg PA
CBHW071537040426
42452CB00008B/1047